THE AIDS BOOKLET

Frank D. Cox

Santa Barbara City College
Sixth Edition

Boston   Burr Ridge, IL   Dubuque, IA   Madison, WI   New York   San Francisco   St. Louis
Bangkok   Bogotá   Caracas   Lisbon   London   Madrid
Mexico City   Milan   New Delhi   Seoul   Singapore   Sydney   Taipei   Toronto

McGraw-Hill Higher Education

A Division of The **McGraw-Hill** Companies

THE AIDS BOOKLET, SIXTH EDITION

This book is printed on acid-free paper.

5 6 7 8 9 0 DOC/DOC 0 9 8 7 6 5 4 3 2 1

ISBN 0-697-29428-5

Vice president and editorial director: *Kevin T. Kane*
Executive editor: *Vicki Malinee*
Developmental editor: *Tricia R. Musel*
Senior marketing manager: *Pamela S. Cooper*
Editing associate: *Joyce Watters*
Senior production supervisor: *Mary E. Haas*
Coordinator of freelance design: *Rick D. Noel*
Photo research coordinator: *John C. Leland*
Compositor: *Shepherd, Inc.*
Typeface: *10/12 Garamond*
Printer: *R. R. Donnelley & Sons Company/Crawfordsville, IN*

Cover image: *Patricia G. Shields*

www.mhhe.com

I wish to dedicate *The AIDS Booklet* to Jack W. Shields, M.D. He recognized early on the danger of AIDS to both the individual and the culture as a whole. Far sooner than most people, he saw the need to mount an all-out war against this destructive new disease. His constant counsel and review of the material contained herein has been invaluable to the production of *The AIDS Booklet.*

C. AIDS ... Africa and Asia ... 1994
D. Who Are the People with AIDS? 21
E. How HIV/AIDS Affect Women and Children ...

TABLE OF CONTENTS

Foreword   vii
HIV Disease: Food for Thought   ix
Updated AIDS Data   x
Issues and Trends Update   xi

PART **1** *WHAT IS HIV/AIDS?* *1*

A. Self-Test   3
B. Brief History of HIV/AIDS in America   4
C. What Is HIV/AIDS?   6
D. The Apparent Incubation Period and Usual Course of HIV/AIDS   11
    Home Testing for HIV   *11*
E. What Are the General Symptoms?   12
F. What Are the Common Illness Patterns Associated with AIDS?   12

PART **2** *HOW MANY PEOPLE HAVE AIDS? WHERE ARE*
              THEY? WHO ARE THEY? *15*

A. How Many People Have AIDS?   15
B. Where Are the People with AIDS?   15
C. AIDS Cases by Metropolitan Area, to January 1999   19
D. Who Are the People with AIDS?   21
    The Economic Impact of HIV/AIDS   *22*
E. How HIV/AIDS Affect Women and Children   25

PART **3** *HOW DO YOU CONTRACT HIV/AIDS?* *29*

A. High-Risk Behaviors   29
B. Sexual Behavior   30
C. Drug Abuse   33
D. Blood Supply and Transfusion   36
    HIV/AIDS Transmission Via Hollow Steel Needles   *37*
E. Casual Transmission (The Transmission of HIV/AIDS Without Sexual
    Contact or Intravenous Drug Use)   38
    The Impact of HIV/AIDS on Health Professionals   *40*

PART   [4]   *HOW CAN YOU PREVENT AIDS?*   *41*

A.   Establishing and Maintaining Caring and Loving Relationships   41
B.   Specific Precautions Caring Persons Can Take to
     Avoid HIV/AIDS Risk   44
C.   If You Are Sexually Active   45
     The Impact of HIV/AIDS on Confidentiality   *46*
D.   Drug Use   50
E.   Donate to Blood Banks   50
F.   Vaccines, Drugs, and Cures   51
     The Impact of HIV/AIDS on the Law   *54*

Summary: Preventing HIV/AIDS   55
Appendix A: The Centers for Disease Control's Revised Definition of AIDS
            as of January, 1993   57
Appendix B: Ten Points for World AIDS Day: AIDS and the Family   58
Appendix C: Costs of Commonly Used Agents and Laboratory Tests for
            HIV-Infected Adult   61
Appendix D: A Technical Description of HIV-1 Sickness Transmission   64
References   67
Glossary   69

FOREWORD

When fully grown and ready to have children, your body normally contains 100 trillion cells. Twenty-five trillion of these are red blood cells, called erythrocytes. These are produced in the bone marrow to carry oxygen to all other cells and carbon dioxide back to the lungs via the blood circulation. Another 25 trillion are white blood cells, called lymphocytes, which are formed in lymph glands located in many parts of the body. In the lymph glands, large lymphocytes extrude a variety of globular substances, including soluble antibodies and proteins, which affect growth in other cells. Small lymphocytes, derived from the large ones by loss of substance, emigrate from the lymph glands and migrate randomly throughout the body to carry DNA and other molecules that control growth in remaining body cells and provide the latter with protection against many kinds of infections.

Consequently, if the lymphocytes are altered or destroyed, the remaining body cells may suffer from malnutrition, poor control of cell growth, and poor resistance to a wide variety of infections. During altered function we see a variety of cell growth or immunologic disorders. When the lymphocytes, large and small, are mostly destroyed the end results are combinations of weight loss with emaciation; disorderly functions or cancer in lymphocytes or other cells; and, finally, death from a variety of infections. This is exactly what happens when the human immunodeficiency virus type 1 (HIV-1) fouls up the genetic machinery in lymphocytes to produce progressive HIV sickness, whose end-stage is called AIDS (Acquired Immune Deficiency Syndrome).

In 1981 human AIDS surfaced as an unusual disease in young gay men. In 1983–84 the retrovirus which causes AIDS was identified via culture of lymphocytes. During the 1980s HIV spread via infected lymphocytes in blood, semen, uterine cervical secretions, and maternal colostrum to engender the worst communicable disease the human race has experienced in modern times. Seventeen years of intense and well-funded research have not yet provided curative medications or effective vaccines for progressive HIV-1 sickness or any other retroviral disease. Therefore, the focus must be on lymphocytes and prevention.

Jack W. Shields, M.D.; M.S., F.A.C.P.

HIV SICKNESS: FOOD FOR THOUGHT

- HIV sickness is indiscriminate with respect to race, gender, sexual orientation, or occupation.
- Free love is costly in terms of HIVs.
- Love with care prevents HIVs.
- Buying sex buys HIVs.
- Selling sex sells HIVs.
- Trading sex for drugs trades HIVs.
- Circumcision deters HIVs in men.
- Contraceptive barriers deter HIVs transmission.
- Injecting drugs shoots HIVs.
- Sexually transmitted diseases aid HIVs transmission.
- Ignorance breeds and education prevents HIVs.
- Genital cleanliness helps to avoid HIVs.
- Good communication helps HIVs prevention.
- To yourself be true. You will not spread or get HIVs from another.
- Care for others with AIDS like you would like to be cared for yourself.
- You might be next if you don't care or are careless.

Dr. Jack Shields

UPDATED AIDS DATA: CENTERS FOR DISEASE CONTROL HIV/AIDS SURVEILLANCE, JULY 1999*			
Total cases	711,344	New York City	111,870
		San Francisco	26,715
Under 5 years	6,672 (1%)	Los Angeles	39,863
5–12 years	1,924 (0%)	Miami	22,283
13–19 years	3,564 (0%)	Washington, D.C.	20,623
20–29 years	120,773 (17%)	These five cities account	
30–39 years	319,947 (45%)	for 31% of all U.S. cases	
40–49 years	183,195 (26%)		
50 & over	75,266 (11%)		*Adults*
		Males	588,124 (83%)
		Females	114,621 (16%)
		White	311,375 (44%)
Total deaths	420,201	African American	262,317 (37%)
Women	60,299	Hispanic	129,555 (18%)
Men	359,902	Other	7,167 (1%)
Death rate	59%		

*Slight variance in data due to reporting method differences.

Death rates* for leading causes of death among persons aged 25–44 years, by year—United States, 1982–1996[†]

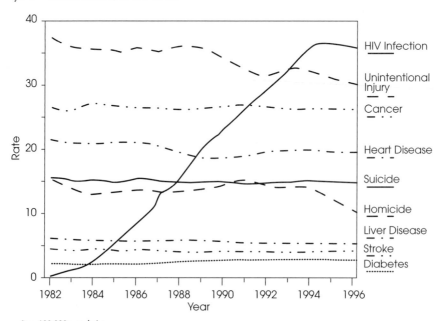

*per 100,000 population.
†Based on underlying cause of death reported on death certificates, using final data for 1982–1995 and preliminary data for 1996.

ISSUES
AND TRENDS UPDATE

1. HIV/AIDS continues to spread in the world. AIDS continues to reduce the average life expectancy in many countries. For example, life expectancy in Zimbabwe has dropped from 65 years to 39 years because of AIDS. In the United States HIV/AIDS is still most often spread via intravenous drug use and male homosexual relations. However, HIV/AIDS spread via heterosexual intercourse has slowly increased.

2. Due to HAART (highly active anti-retroviral therapy) lower death rates in men have been reported for the past two years although the numbers appear to be rising again (JAMA Oct. 6, 1999). HAART, combined with better treatment of opportunistic infections, does prolong HIV infected persons' lives. Unfortunately, this has led to increased public complacency about HIV/AIDS despite the fact that such treatment does not cure those infected with HIV/AIDS. A STOP AIDS survey of 21,857 gay men in San Francisco found 39 percent reporting unprotected anal sex in 1997, up from 30 percent in 1994 (Associated Press. January 31, 1999 as reported in the *Santa Barbara News Press*, A6).

3. Currently HAART helps about 50 percent of patients and out of these, about 30 to 40 percent experience negative side effects, such as diabetes, cholesterol problems, coronary artery disease, kidney stones, neuritis, unsightly fat deposits on the back, shoulders, and abdomen and weak, spindly arms and legs (*NEJM*. October 29, 1998: 1296–1297).

4. At a cost of $10 to $15 thousand annually, few can afford HAART treatment without subsidy in the United States or elsewhere. Treatment dramatically reduces virus RNA load in blood, but does not eliminate provirus mutation. More importantly, treatment does not eliminate provirus-infected lymphocytes from the lymph glands or small motile lymphocytes called "CD^4 memory T-cells" migrating in blood and ejaculated semen (*NEJM* 1998: 3391846). Poor compliance with complicated (seventy pills or more a week. *Lancet.* December 1998) treatment schedules (*APA Monitor.* October 1998) and emergence of drug resistant forms are major problems.

5. When HAART treatment stops, whether it be after two weeks or three years, the virus load returns, often with drug resistance (*NEJM*. October 29, 1998: 1319–1320). Ten new anti-retroviral drugs were FDA-approved for use in HAART programs last year. So far, none have proved more effective than the combinations first used. Injections of Interleukin-1, a soluble

growth promoting factor released by cytoplasmic shedding from healthy lymphocytes in tissue culture (see fig. 1.3), showed some promise in studies reported from the National Institutes of Health. However it has failed to eliminate infected lymphocytes (JAMA, Oct. 13, 1999).

6. It was hoped that the reduction in HIV via HAART could be maintained by reducing the drug intake necessary initially. Unfortunately HIV load increases with reduced maintenance strategies (*NEJM.* October 29, 1998; 1269, 1319–1320).

7. During and after treatment, HAART does not significantly reduce the numbers of HIV-infected lymphocytes in lymph glands or the numbers of lymphocytes containing random-integrated HIV-1 RNA in the form of integrated proviral DNA in the nuclei of small lymphocytes migrating in blood, semen, uterine cervical secretions, and maternal milk. Thus, HAART does little to prevent person-to-person spread of HIV/AIDS and may lead to a false sense of security on the part of the general public (*JAMA* 1998: 279: 641–642; *JAMA* August 19, 1998, 587).

8. Effective vaccines for prevention still remain doubtful (see p. 51) (*The Scientist.* September 28, 1998: 1, 6, 15; *Lancet.* October 24, 1998: 1323). The main problem with respect to cures is the capacity of HIV-1 to mutate in lymph glands, such that many forms of sickness evolve and resistance to HAART develops rapidly (Lancet, Aug. 28, 1999 and NEJM, Sept. 30, 1999).

9. "Harm reduction" for intravenous drug users seems to help control HIV/AIDS infection in countries such as Holland and Switzerland. This includes programs such as needle exchange, methadone treatment, medical prescription of drugs, and advertising risk reduction programs among intravenous drug users. The United States' "war on drugs" policies preclude many such programs.

10. The original animal source of HIV-1 was genetically traced via polymerase chain reaction (PCR) and DNA-sequencing to a species of chimpanzee in central Africa (*Nature* 1999: 397, 436). These chimps, seemingly resistant to AIDS, have been used as meat by Africans and few remain for study.

11. Some infected humans are long-term survivors of HIV/AIDS. Possibly they were born with resistant genes. Studies of such "immune" persons continue but are not likely to help much until reliable systems of genetic engineering become practical in humans.

12. The use of zidovudine (AZT) during late pregnancy to prevent HIV/AIDS in offspring has had notable success (*NEJM.* April 1999: 1040–1043). However, this cures neither parent meaning that most of the children born of infected mothers will lose their parents within 5 to 10 years. The newest development is Nevirapine. A single dose given to the mother during labor and to the newborn within 72 hours after delivery it may reduce the chances of HIV infection (Lancet 1999, 354; 795).

13. Ceasarean section birth also reduces the risk of fetal infection, however, it puts the HIV+ mother at more risk (*NEJM.* April, 1999: 977–987, 1032–1033).

14. The XII International HIV/AIDS conference in Geneva (June 29–July 12, 1998) concluded that the newest treatments are not working as well as hoped in developed nations, and are not reaching the millions of HIV infected people in poor countries (*Science.* July 10, 1998: 159–160).

15. Many countries sample test their populations for HIV infection. In the United States only those with full blown AIDS are reported. Thus the extent of the epidemic in the United States is underrepresented since it does not include an accurate count of the number of HIV+ persons. Politicalization of HIV/AIDS has harmed public health attempts to control the epidemic. Such standard actions as sampling the population for infection and partner tracing, identification, and notification have been denied. This contributes to the spread of HIV/AIDS.

16. Education and prevention remain our most important weapons in the war against HIV/AIDS. The proposed 1999 U.S. budget for governmental spending on HIV/AIDS is $1.73 billion–$573 million for etiology and pathogenesis; $482 million for treatment; $180 million for vaccine development; and an untold sum for administration. Unfortunately this leaves little for education and prevention work. Since HIV/AIDS can be controlled via one's personal behavior, the personal decision to avoid risky behavior remains our most effective deterrent against the epidemic. Engaging in behavior known to risk HIV/AIDS infection while expecting science and medicine to protect one is totally unrealistic and actually contributes to the epidemic.

1

WHAT IS HIV/AIDS?

Recent headlines reporting the drop in new AIDS cases may lead you to believe that the HIV/AIDS epidemic is now under control. Nothing could be further from the truth. Despite some good news, the threat of HIV/AIDS is alive and well. The it-can't-happen-to-me attitude, combined with optimistic reports suggesting the defeat of HIV/AIDS, is a dangerous combination. It may, in fact, lead one to careless instead of preventative behaviors.

Over the past years you have read and heard a great deal about HIV/AIDS. In all probability, your friends and you have discussed HIV/AIDS and what can be done to protect yourselves from contracting it. The federal, state, and local governments as well as private organizations are actively educating the public about HIV/AIDS. The reason for all of the publicity and concern about HIV/AIDS is the fact that AIDS kills people, and presently there is no cure. In 1993, AIDS-related deaths were the eighth leading cause of death in the United States. By 1996 the Centers for Disease Control (CDC) reported AIDS as the sixth leading cause of *premature* death following accidents, cancer, heart disease, murder, and suicide. AIDS is the leading cause of death among men aged 25 to 44 and among women 35 to 44 years of age. Table 1.1 and figure 1.1 will give you an overview of just how dangerous AIDS is.

The Secretary of Health and Human Services issued the following statement to commemorate the ninth annual World AIDS Day (December 1, 1996):
"Today around the world 8,500 of our fellow humans will become newly infected with HIV. Another 8,500 people will be infected tomorrow, the next day, and the next day. . . ."

Table 1.1		
REPORTED AIDS CASES AND AIDS DEATHS ANNUALLY IN THE UNITED STATES (ADULTS)		
Year	*AIDS Cases*	*Deaths Due to AIDS*
1995	74,180	48,979
1996	67,943	42,155
1997	59,657	28,588
1998	48,268	20,200

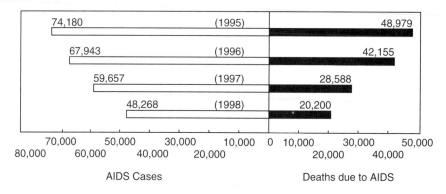

Figure 1.1   AIDS Cases and Deaths Reported Annually

"In the United States alone, at least 40,000 Americans become infected with HIV every year. Every day, over 100 Americans die of AIDS-related complications. These are our sons and daughters, our fathers and mothers, our friends, lovers, and loved ones."

"While the powerful combination therapies offer hope to those who can access them, we must refrain from premature declarations of victory. The AIDS pandemic is not over."

By now you probably know that HIV/AIDS has to do with the most intimate aspects of your life, namely sexual behavior. Although HIV is tied closely to the sexual aspects of your life, it is also related to drug use, blood transfusion, pregnancy, and birth. At present there are no effective vaccines against HIV/AIDS. Your best protection against this deadly disease is establishing caring, respectful relationships and educating yourself so that you can avoid behaviors that expose you to infection.

PREVENTION! EDUCATION! PREVENTION! Avoid unsafe sex practices; avoid intravenous drug use. This advice has become the formula for surviving in the world of AIDS. Controlling your own behavior can help prevent human immunodeficiency virus (HIV) infection. It has been almost 16 years since Luc Montagnier of the Pasteur Institute (in France) published the first report on the virus (HIV) that is now known to cause AIDS. Despite the high-powered arsenal of contemporary biology and media hype, there is nothing on the horizon resembling a cure for AIDS. Nor is science anywhere near a workable vaccine against HIV.

Although you read about different medicines that help people with AIDS, these act only to prolong life rather than to cure the disease.

Since the recognition of HIV/AIDS, the emphasis has been on research to cure the disease in infected people. To date this has failed. The message today is: **THE ONLY DEFENSE AGAINST HIV/AIDS IS PREVENTION.** To ignore the risks is now suicidal because of the large number of people infected with HIV.

Because of the long apparent latent period after HIV infection before symptoms appear, few teenagers actually have AIDS. This may lead you to believe teens cannot contract the disease. However, recent figures clearly show that many teenagers are infected. For example, among men aged 25 to 44, AIDS accounts for 16.5% of all deaths and among women, 4.8% of all deaths. When one considers that HIV sickness averages around 12 years before actual AIDS sickness occurs, it is obvious that many of those who die in this age group contracted the virus while in their teens. If you remain convinced that you are not at risk for AIDS as a teenager, consider this: As of July 1997, there were approximately 3,000 cases of AIDS in 13- to 19-year-olds, a 100-fold increase from the number in 1983. In the same period, adult cases increased about 55-fold. Thirty-seven percent of these cases were female.

HIV is spread quickly and silently throughout the teen-age community along with a rise in other sexually transmitted diseases. Infection can be prevented if each person avoids risky sexual- and drug-related behaviors. With better education about HIV, self-control, increased care, and respect for one another, the spread of HIV can be contained.

How can you live a full life but avoid contracting or spreading HIV? In order to find the answer to this question, you must know what HIV infection is, how it progresses into AIDS, how HIV is transmitted, and what steps you can take to prevent its spread. Since the discussion of HIV/AIDS involves intimate behaviors, the answers involve your moral and ethical judgments, both about yourself and your relationships with others. This booklet will help you make decisions about ways to live your life and interact with others in order to minimize your chances of contracting or spreading HIV.

You will find many medical terms that are new to you. We have placed such terms in ***boldface italic*** and have given a short definition. You will also find a listing of all such terms in the glossary. **It is *not* important that you learn all of these terms, but understand that HIV infection, resulting in AIDS, is a complicated, dangerous, life-threatening health problem.**

A. SELF-TEST

Answer the following questions true or false.

1. HIV is a sexually transmitted disease.
2. Presently in the United States, AIDS mostly strikes men who engage in homosexual activity and people who use injectable drugs (using needles to shoot drugs directly into the veins).
3. Although AIDS infects more men than women, during the past few years the percentage increase of women contracting AIDS has been higher than the percentage increase of men.
4. People can be infected with the AIDS virus (HIV) without knowing it and without having symptoms of the disease.

5. People infected with HIV who do not show AIDS symptoms can infect others.
6. In many parts of the world, as many women and children as men are HIV positive.
7. Intravenous drug users who share needles can expose themselves to HIV.
8. Abstinence (avoidance of sexual relations) is the most effective way to prevent the transmission of HIV by sexual contact.
9. The chances of becoming infected with HIV through a blood transfusion are greatly reduced because HIV screening tests are now used to screen blood donations.
10. A person cannot get HIV/AIDS from donating blood.
11. An HIV-positive pregnant woman can transmit the virus to her unborn child.
12. An infected mother can transmit HIV to her newborn via breast milk when nursing the child.
13. HIV is most often transmitted through blood or semen.
14. HIV cannot be transmitted through casual contact such as shaking hands or eating with an infected person.
15. The proper use of latex condoms during sexual intercourse can help prevent the transmission of HIV.
16. A person practicing sexual abstinence and who refrains from injecting drugs has little chance of becoming infected with HIV.
17. HIV weakens a person's immune system, making him or her susceptible to many different diseases.
18. HIV alone does not kill a person.
19. People with AIDS die from other diseases that they acquire because of their weakened immune system.
20. Most people infected with HIV will sooner or later develop AIDS or a related fatal disease.

(In light of current knowledge, the answers to all of these questions are true.)

B. BRIEF HISTORY OF HIV/AIDS IN AMERICA

1976–1979   Unrecognized cases of AIDS in humans start to appear in the United States. It is not known exactly how the AIDS virus initially entered the human population. Theories range from transmission via monkey bites in Central Africa to contaminated vaccines. You will hear and read about many such theories, but none have yet been proven.

1980   Fifty-five young men are diagnosed with similar infections apparently linked to some new virus that will later be identified as HIV.

June 5, 1981   The Centers for Disease Control (CDC) publishes the first report on a new epidemic based on several Los Angeles cases. The report is published in the *Morbidity and Mortality Weekly Report* and is entitled *Pneumocystis Pneumonia,* which remains a major killer of people with AIDS.

1981–1985   AIDS is mostly limited to White homosexual men.

1983	National Institute of Health and Pasteur Institute in France identify a retrovirus as the likely cause of AIDS.
April 23, 1984	Discovery of HIV is announced by the Department of Health. French and American researchers debate over who should be credited with the discovery when it becomes public.
September 1984	HIV is found to infect cells in the brain and central nervous system.
March 2, 1985	First AIDS antibody tests for the screening of blood are publicly released.
April 1985	First International Conference on AIDS held in Atlanta, Georgia. The full, worldwide scope of the AIDS epidemic is disclosed. First large-scale media coverage given to AIDS.
October 1985	Movie star Rock Hudson dies from AIDS. This brought the American public's attention to the deadly nature of the disease and is considered a turning point in arousing public awareness of AIDS.
April 1986	Second International Conference on AIDS is held in Washington, D.C. The first treatment to extend the lives of people with AIDS (AZT, see p. 52) is publicized. AIDS is moving into populations using drugs and minorities. AIDS starts to threaten the general population.
June 1990	Sixth International Conference on AIDS held in San Francisco. AIDS becomes politicized and there are demonstrations and disruptions by various groups of activists.
January 1991	100,000th death from AIDS.
July 1991	200,000th diagnosed case of AIDS.
Nov. 7, 1991	Basketball great Magic Johnson announces he is HIV positive, spurring greater public awareness of HIV transmission among heterosexual people.
Feb. 6, 1993	Tennis star Arthur Ashe dies from AIDS contracted through blood transfusion during heart surgery in 1983. Publicity surrounding his death helps to create greater public awareness.
June 1993	The Ninth International Conference on AIDS concludes that there is nothing on the horizon resembling a cure for AIDS, nor is there anything like a workable vaccine. Prevention through education remains the only weapon in the fight against AIDS.
May 1993	300,000th case of recognized AIDS. Polymerase chain reaction (PCR) technology shows germinal centers in lymph glands are the most common primary targets of HIV.
September 1993	200,000th death from AIDS.
June 1994	Tenth International Conference on AIDS concludes that so little progress is being made that future conferences will be held every other year rather than every year. Conference reports HIV-positive American women and women with AIDS are increasing more rapidly than in the past. The 400,000th person is

	diagnosed with AIDS. Concorde trials reveal that AZT does not significantly alter the natural course of AIDS in adults.
January 1995	Greg Louganis, Olympic diving champion, publicly announces he is HIV positive.
February 1995	Several studies change our understanding of the long latency period after HIV infection and before AIDS appears. It is, in fact, not a true latency period at all as the virus is actively invading lymphocyte cells during the time that there are no overt symptoms. This leads to the idea of **HIV disease** as a progressive sickness that eventually culminates in AIDS.
July 1996	At the Eleventh International Conference on AIDS, protease inhibitors and other drugs given in varying combinations arouse hope that a breakthrough in treatment of HIV/AIDS is near.
February 1997	The United States experiences the first drop in AIDS-related deaths since the epidemic began.
September 1997	The number of reported AIDS cases in 1996 drops for the first time since the epidemic began.

C. WHAT IS HIV/AIDS?

Confusion persists concerning the cause, spread, and treatment of HIV/AIDS. *Acquired Immunodeficiency Syndrome (AIDS)* is a variable disease caused by a retrovirus that infects lymph glands and destroys lymphocytes through gene alteration; spreading the disease between individuals mostly through semen, blood, uterine secretions, and to a lesser extent, through the placenta and infected mother's milk.

The disease is divided into four stages:

1. *Prodrome* (premonitory symptom of a disease). The time during which aches, fever, and headache are the most common symptoms.
2. *Latency.* When overt symptoms are absent, the infection persists in lymph glands and spreads via small lymphocytes migrating from these glands.
3. *Generalized lymph gland enlargement and/or autoimmune diseases,* such as kidney or bone marrow failure occur.
4. *AIDS.* The body will experience wasting, poor resistance to all kinds of infections; various forms of cancer and lymph gland destruction occur.

Let us break down the acronym AIDS—*Acquired Immune Deficiency Syndrome. Acquired* means the conditions are not inherited but are acquired from environmental factors such as virus infections. *Immune Deficiency* means that the viruses gradually cause deficient immunity, which reflects poor nutrition and low resistance to infections and cancers. *Syndrome* means the viruses cause several kinds of diseases, each with characteristic signs and symptoms. Because the infectious diseases caused by HIV have so many variable manifestations before AIDS appears, **HIV disease** is a good descriptive term to use.

Unfortunately AIDS appears to have a number of meanings. In addition, the term has become highly politicized. For many people, AIDS is an acquired sickness caused by HIV that ends in death from immune deficiency. To others, AIDS means having one or more well-defined diseases, such as lymphomas (cancer of the lymphocytes) or a wasting away of the entire body (wasting or slim disease) and being HIV+. To a few, AIDS and being HIV+ are the same thing.

The virus that causes AIDS has different names, but the term preferred by most scientists is **HIV (human immunodeficiency virus). Viruses** are disease-causing agents and are too small to see with an ordinary microscope. Viruses depend on the living host cells (in the case of AIDS, the lymph cells) to grow and survive.

Although we usually speak of one AIDS virus, there are at least two viruses that can cause AIDS, AIDS-related conditions, and cancers in human beings. HIV-1 is still the most common cause of AIDS worldwide, except in West Africa where HIV-2 is relatively common. HIV-2 appears to be less virulent than HIV-1. In addition, scientists can identify up to nine major genetic subtypes of HIV-1. It should be noted that HIV is a retrovirus—this class of viruses lack DNA and, therefore, depend on the DNA in other bodily cells (in this case, lymphocytes) to reproduce.

People afflicted with AIDS usually suffer from various combinations of severe weight loss, many different types of infections, and several different kinds of cancer. They go through long, miserable illnesses that end in death several years after the initial diagnosis of AIDS. However, this general description does not apply to every person with the disease. Some people with AIDS alternate between periods of sickness and periods of fairly good health, at least early after diagnosis. Other people die within a few months, while a few have lived ten or more years after the initial diagnosis. It is important to understand that having the HIV in one's blood is not the same as having AIDS. Most people carrying HIV in their blood remain symptom-free for several years (*during which time they can infect others, however*). Although most HIV+ people will sooner or later develop AIDS, it is unclear at this time whether all such carriers will go on to develop the disease. To date, approximately 5 percent of HIV+ people have remained free from progressive disease and maintain normal cell counts for a decade or more.

There have even been a few AIDS cases reported in which the person is not HIV+. This has led a few scientists to believe that HIV may not be the only cause of AIDS, although this view is not widely held.

Varying symptoms are associated with AIDS. The reason for this is (as the name implies) that HIV impairs the body's ability to fight infection. It does this by destroying lymphocytes. **Lymphocyte** means "lymph cell" (see figure 1.2).

Lymphocytes feed other cells, control cell growth, and guard against infection. Lymphocytes are the most common kind of cell in our biological defense system, the **immune system.** They help prevent cancers by controlling cell growth, and they help protect against infections by producing **antibodies** (proteins that fight infection).

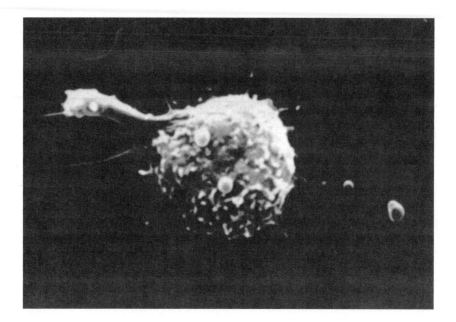

Figure 1.2 A Normal, Medium-Sized Lymphocyte Migrating (Scanning Electron Microscope). Note the tiny globules extruding from the surface. Credit: Jack W. Shields, M.D.

Unfortunately, as the infected lymphocytes produce these antibodies, HIV is also reproducing. Thus, a person with AIDS appears undernourished and wasted, often has cancer, and is not protected from infections. In the media and in the Centers for Disease Control (CDC) definition of AIDS you will come across the name CD^4+ or T-helper cell. This is a lymphocyte in an advanced stage of maturity. Most circulating CD^4+ lymphocytes disappear with the onset of AIDS, along with the germinal centers from which such lymphocytes are produced in various lymph glands. Thus, the CDC suggests that one sign of AIDS is a depressed count of these cells.

Each day you come into contact with many kinds of infectious diseases, but your immune system protects you from getting sick. When you do get an infection such as the chicken pox, the immune system manufactures antibodies that help fight the infection. When you get well, your body usually becomes immune to that particular infection. This is called *acquired immunity,* which means that you will normally not get the disease again. Unfortunately, people do not develop effective acquired immunity to HIV because it grows in the very cells that produce antibodies.

If the lymphocytes in your immune system are damaged or destroyed, as with HIV sickness, the immune system does not respond properly. You are then more susceptible to some of the many infections and diseases that exist within the environment. For example, TB (tuberculosis) has been increasing, and an estimated 10 million Americans now carry the infection. HIV+ people are especially susceptible because of their deficient immune system. The *Journal of the American Medical Association* (August 1994) reported an HIV+ person was 30 times more likely in any given year

Figure 1.3   Lymphocytes under scanning electron microscopy in PHA-stimulated tissue culture (Magnification 10,000 times and 20,000 times in Fig. 1.4): A healthy large lymphocyte extruding many surface globules (arrows), which normally dissolve to become a variety of circulating globulins, including lymphokines, antibodies, and nutrients. The large lymphocytes normally transform into medium-sized lymphocytes through extruding such globules and accompanied by condensation of the nucleus. Credit: Dr. Jack W. Shields.

to become sick with TB. The increase in TB is particularly frightening because some of the newly appearing TB seems to be resistant to drugs that were used earlier to control it. A study done in New York City found that 33 percent of TB cases were resistant to at least one drug, while 19 percent of the cases were resistant to two or more drugs. Unlike HIV, TB is easily spread throughout the population via coughing and sneezing.

Without a healthy immune system you would be sick all of the time. You would also be likely to develop ***malignant*** (cancerous) growths of cells within your body. Thus, a damaged immune system fails to battle cancers that frequently invade people with AIDS.

It should be emphasized that people do not die directly of the causative virus infection, but rather from one of the many diseases, infections, or cancers that develop because of a weakened immune system.

The virus progressively damages and destroys the lymphocytes on which it depends to survive. It is progressive lymphocyte destruction that causes the problems rather than any free circulating HIV in the blood. Over the long run, HIV causes changes in the infected cells that induce cell (lymphocyte) death. As HIV illness progresses to AIDS, people usually contract infections that are normally prevented by healthy antibody-producing and migrating white blood cells. We call such infections ***opportunistic infections*** because they take advantage of the damaged immune system. Unfortunately, modern medicine has yet to find a way to completely repair the immune system once it has been damaged by HIV.

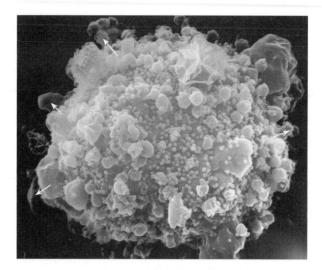

Figure 1.4   An HIV-infected cell extruding many surface globules (arrows) and hundreds of tiny, round, pox-like particles, which are HIV. During the apparently latent phases of HIV sickness, most of the tiny virus particles are precipitated by antibodies extruded from the large lymphocytes. As a result, few HIV are found in circulation, as opposed to small lymphocytes carrying retrovirus in their nuclear DNA. The virus particles themselves are virtually harmless to other cells. Eventually, AIDS sets in, when most of the large lymphocytes in the germinal centers of lymph glands are destroyed. This leads to a gradual reduction in the number of circulating lymphocytes. Credit: Dr. David Hockley, EMP, NIBSC and Science Photo Library, U.K.

Viruses can be divided into: (1) DNA viruses containing genes that direct virus growth, and (2) RNA viruses lacking DNA that partially depend on the genes inside of other cells for growth and reproduction of the virus. HIV-1 is an RNA virus, also called a retrovirus, because it randomly reverse transcribes and inserts RNA into the DNA of the host cells—in this case lymphocytes—which in turn function abnormally or are killed.

As mentioned earlier, these white blood cells, the lymphocytes, are responsible for helping to defend the body against infection. However, by infecting the lymphocytes, HIV sabotages their ability to function properly. Thus, the very cells that protect and help the body to fight infection not only fail to function properly, but also help to spread the virus. Anything that disturbs immune function such as another disease activates these cells to fight the intruder, but this also increases production of HIV.

Lymphocytes are the most common migrating cells in the body, constituting 1 to 2 percent of body mass in healthy humans. They reproduce at such a rapid rate that all of the cells can be replaced by the body within a 48-hour period. They migrate rapidly throughout the body. As a result, if lymphocytes become infected with HIV, the virus is quickly carried throughout the body to infect other cells or is carried via semen, blood, and mother's milk to infect other people.

> ### HOME TESTING FOR HIV
>
> *T*he testing system is comprised of three integrated components: an over-the-counter home blood collection kit, HIV-1 antibody testing at a certified lab, and a test result center that provides the results, counseling, and anonymous referrals. There are a number of problems with HIV home testing that need to be addressed: The lack of face-to-face counseling when the results of such tests are reported is a major concern of health-care workers. This is why it is important to visit a testing center to discuss the results rather than receiving the results by telephone only. Other things that worry HIV/AIDS workers about home testing include: Will people read and follow the directions? Is the mail system really that reliable, and will it avoid damaging the package? Will the person taking the home test put an adequate amount of blood on the card to be sent? Any person who finds the code number of the person being tested could call for the results, thus compromising confidentiality.

D. THE APPARENT INCUBATION PERIOD AND USUAL COURSE OF HIV/AIDS

Within about three to six weeks after first exposure, some HIV+ persons develop a 7- to 21-day illness with enlargement of the lymph glands, sore throat, fever, muscle aches, headache, and a skin rash that, in some cases, looks like measles. HIV can be detected in circulating blood lymphocytes at this time, but tests for antibodies to HIV seldom become positive until six weeks to six months later. This early form of illness usually disappears or often is so mild that it is not even remembered. *However, the infected person is now contagious for the remainder of his or her lifetime and can transmit HIV to other persons.* There is also evidence that HIV becomes more infectious to others as time passes. Since a person does not know if he or she is HIV+ without a test (during the early years of infection), the FDA (Food and Drug Administration) has approved home testing kits for HIV.

A few people develop brain infections severe enough to require hospitalization during the early stage of HIV infection. The usual signs are severe headache, drowsiness, pain in the eyes when looking at a bright light, fever, stiff neck, painful muscles, and a general state of collapse. This painful form of viral encephalitis or meningitis is usually transient and goes away without special treatment.

Subsequently, months or years may pass without any overt symptoms in an infected person. However, during this time HIV is being actively produced thereby weakening the immune system. Although this time period is called the incubation or latency period, it is really not clinical latency at all. Rather it is a period during which replication of the virus in the lymphatic system continues, but there are no overt symptoms of the disease. Hence, it is an apparent latency period rather than a true latency period. **It is most realistic to think of this process as HIV sickness, a progressive sickness that ultimately culminates in AIDS, when multiple opportunistic diseases eventually cause death.** The precise period of HIV sickness before the development of AIDS for any individual is presently unknown. Some researchers are predicting that this period may be as long as twenty years. Be that as it may, once AIDS appears, death will usually follow within the next few years, even though improved forms of treatment are prolonging survival.

E. WHAT ARE THE GENERAL SYMPTOMS?

At least 80 percent of those infected with HIV will eventually die from AIDS or an AIDS-related condition (many researchers suggest that over enough time 100 percent of such persons will die from the sickness).

As HIV sickness develops, the following symptoms are likely to appear, singly or together.

1. Loss of appetite with weight loss of ten or more pounds in two months or less.
2. Swollen glands (lymph glands) in the neck, armpits, or groin that persist for three months or more.
3. Severe fatigue (not related to exercise or drug use).
4. Unexplained persistent or recurrent fevers often with night sweats.
5. Persistent unexplained cough (not from smoking, cold, or flu) often associated with a shortness of breath.
6. Unexplained persistent diarrhea.
7. Persistent white coating or spots inside the mouth or throat that may be accompanied by soreness and difficulty in swallowing.
8. Newly appearing persistent purple or brown lumps or spots on the skin. On White people, they look like small bruises; on African-American people, the spots appear darker than the surrounding skin.
9. Nervous system impairment including general dementia, loss of memory, inability to think clearly, loss of judgment, and/or depression. Other problems such as headaches, stiff neck, and numbness or muscle weakness may occur.

Any of these symptoms may be caused by diseases other than AIDS, and this makes self-diagnosis difficult. However, if such symptoms persist or several appear at the same time, you should suspect exposure to HIV and should immediately see a physician familiar with the disease.

F. WHAT ARE THE COMMON ILLNESS PATTERNS ASSOCIATED WITH AIDS?

The following illnesses have names that may not be familiar to you because such illnesses have been rare and unusual in the past. You do not need to learn the names, but it is important to understand that the illnesses associated with AIDS are serious and life threatening. Doctors define some of these illnesses as follows:

1. ***Wasting or Slim Disease*** is severe weight loss, body wasting, and weakness often associated with chronic diarrhea and persistent coughing.
2. ***Kaposi's Sarcoma*** (**KS**) is an unusual cancer of blood vessels that occurs in the skin, mouth, lungs, liver, lymph glands, and so forth, giving rise to purplish spots or tumors. This is seen most often in homosexual men, although the prevalence of KS appears to be dropping.
3. ***Lymphomas*** are cancers of the lymphocytes. Lymphomas affect 3–10 percent of AIDS patients and are especially likely to occur in the brain.
4. ***Pneumocystis Carinii Pneumonia*** (**PCP**) is the most common opportunistic infection associated with AIDS in the United States. It is the cause of death in 60 percent of those persons who have already died from AIDS. PCP is due to a tiny parasite that infects the lungs, causing infected persons to slowly smother.
5. ***Aids-Related Dementia*** (**ARD**) is a degeneration of the brain and spinal cord leading to the nervous system impairments previously described.
6. Women commonly show a high incidence of *vaginal fungus infections, uterine cervical dysplasia* (abnormal growth of the cervix), and ***endocervical cancer.***

A variety of other illnesses involving skin, muscles, heart, and endocrine glands are prone to occur during the course of AIDS. However, the illnesses listed above are the most common ones accompanying AIDS in the United States. Eventual death from AIDS is a slow, long-term process during which time the infected person may suffer through several of the illnesses described. Some people may have several or all of the illnesses, one after the other or sometimes in combination. It usually takes several years after the diagnosis of AIDS for infected people to die. Although there are periods of partial or complete disappearance of symptoms, usually temporary, early in the course of the disease, such periods become shorter and rarer because infections appear more frequently as the body's lymphocytes become disabled. Thus, the person with AIDS is in and out of the hospital, often unable to work or live a normal life for any length of time, especially in the later stages of the disease.

Before AIDS and the symptoms appear, we may see a variety of sicknesses wherein the symptoms are caused by disturbances of antibody production. Persistent generalized lymph gland enlargement (PGL) is often seen, and the enlarged glands can be felt in the neck, armpits, and groin. Unusual reactions to medications and/or obscure fevers can also accompany PGL.

HIV-related sickness can also cause antibodies to precipitate in the kidneys, leading to kidney failure. A wide variety of skin disorders are common. Another manifestation is bleeding into the skin, owing to the lack of blood platelets. Also there may be general weakness, owing to anemia caused by lack of red blood cells. Such forms of HIV sickness are actually not diagnosed as AIDS because there is no shortage of circulating lymphocytes and the $CD^{4}+$ cells usually number more than 200 per micro liter (see CDC definition, appendix A; a micro liter is 1/100,000 liter).

HOW MANY PEOPLE HAVE AIDS?
WHERE ARE THEY? WHO ARE THEY?

A. HOW MANY PEOPLE HAVE AIDS?

Approximately 2–3 million cases of AIDS have been reported to The World Health Organization (WHO) as of 1999. Due to under-diagnosis and incomplete report-ing, WHO estimates that 9 million AIDS cases have occurred since the beginning of the pandemic. In many nations, the number of people reported with AIDS is dou-bling every 6–12 months.

This doubling rate has slowed down in the United States, where the disease seems to be stabilizing at about 50,000 to 60,000 new AIDS cases per year. The rate of increase is much slower among homosexual men and blood transfusion recipients, but it is increasing among intravenous drug users, women and children, African-Americans, Hispanics, adolescents, and heterosexuals. As of July 1999, there were 711,344 reported cases of AIDS in the United States: 420,201 people (59 percent) diagnosed with AIDS have died.

The Centers for Disease Control, or CDC (the official U.S. government public health bureau responsible for tracking disease), in Atlanta, Georgia, predicts approx-imately 750,000 total American cases by January 1, 2000 (figure 2.1).

It is hard to imagine, but at this time it is estimated that there are 1–2 million peo-ple in the United States who have already been exposed to the disease, and this is only a guess as universal testing for HIV is not done. Being exposed to HIV means that the virus is in a person's blood, but there are no symptoms of the disease. Such a person is termed seropositive (HIV+). *A virus carrier can, however, spread the disease.* We do not yet know how many HIV+ people will actually develop AIDS. Estimates suggest that from 70–100 percent of those carrying HIV will sooner or later develop AIDS. Many, if not most, of these people do not know that they are infected and, therefore, do not realize they can spread the disease to others. Fortunately, inexpensive blood tests are now available to uncover possible HIV infection. The tests are administered at various test sites throughout the country and by most doctors. You can call your county med-ical health department to find the location of the test center nearest you.

B. WHERE ARE THE PEOPLE WITH AIDS?

All states now report some people with AIDS, although the numbers vary greatly. Two states, New York and California, account for 35 percent of all reported AIDS

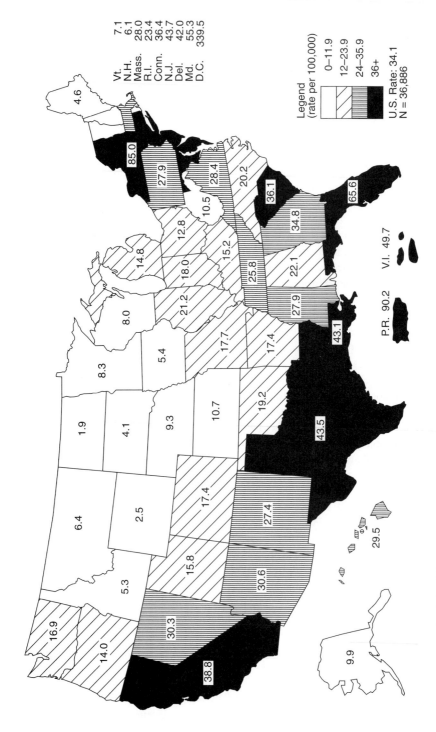

Legend
(rate per 100,000)

0–11.9
12–23.9
24–35.9
36+

U.S. Rate: 34.1
N = 36,886

Vt.   7.1
N.H.   6.1
Mass. 28.0
R.I.   23.4
Conn. 36.4
N.J.   43.7
Del.   42.0
Md.   55.3
D.C. 339.5

P.R. 90.2   V.I. 49.7

Figure 2.1   Male adult/adolescent annual AIDS rates per 100,000 population, for cases reported in 1998, United States.

cases (figure 2.1). People with AIDS are concentrated in New York City, San Francisco, Los Angeles, Washington, D.C., and Miami. It is also within Los Angeles and San Francisco that AIDS was first officially reported in 1981.

It has been suggested that, following the successful example of Australia, the areas where AIDS is highest should be targeted rather than spreading prevention efforts evenly throughout the country. Unfortunately, in the United States the politicalization of AIDS makes it difficult to target areas or groups. Targeted groups may claim discrimination even though AIDS is spreading more quickly among them.

Approximately 208 countries report people with AIDS. For some countries the accuracy of the reports is unclear. Many experts feel that there are more HIV+ persons and persons sick with AIDS in the world than current reports show (see figure 2.2).

There appear to be three major patterns of HIV transmission worldwide. Pattern I is seen in North America and Western Europe. In these areas, the common population groups affected are homosexual men and injecting drug users. HIV/AIDS is found in six men for every one woman. Pattern II areas are sub-Saharan Africa and some parts of the Caribbean. In these areas, HIV/AIDS is commonly found in heterosexuals, and as many women as men are infected. Pattern III includes Asia, Eastern Europe, North Africa, and some of the Pacific countries where infection rates are still low but increasing. In these areas, there is no clear predominant mode of transmission. Globally, heterosexual transmissions outnumber all other transmission categories.

In Eastern Europe the reuse of blood-contaminated needles has been a common method of spreading HIV. The steepest increases now being seen are in South and Southeast Asia, especially Thailand where heterosexual infection by means of prostitutes is widespread.

The numbers reported by the CDC are always less than actual figures because each case is evaluated in detail before being added to the published statistics. If there is any reasonable doubt, owing to incomplete records or lack of crucial information after the death of a person with AIDS, the case may not be recorded until the doubt is cleared up. Because of the time lag with the CDC publication of AIDS statistics, the current available numbers are always two months to a year behind actual numbers of cases and deaths.

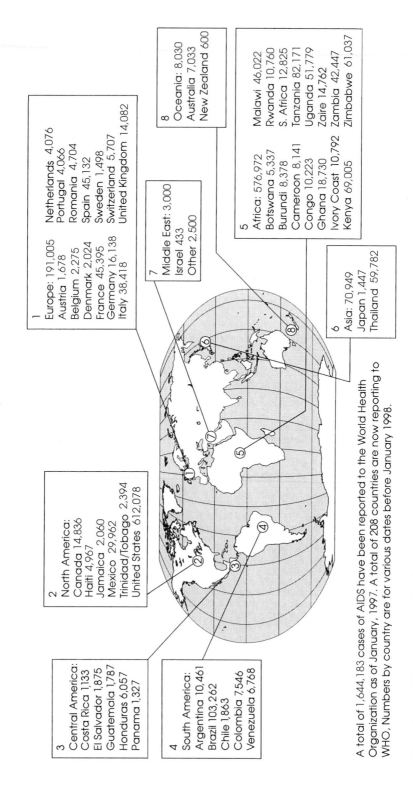

1

Europe: 191,005
Austria 1,678
Belgium 2,275
Denmark 2,024
France 45,395
Germany 16,138
Italy 38,418
Netherlands 4,076
Portugal 4,066
Romania 4,704
Spain 45,132
Sweden 1,498
Switzerland 5,707
United Kingdom 14,082

2

North America:
Canada 14,836
Haiti 4,967
Jamaica 2,060
Mexico 29,962
Trinidad/Tobago 2,394
United States 612,078

3

Central America:
Costa Rica 1,133
El Salvador 1,875
Guatemala 1,787
Honduras 6,057
Panama 1,327

4

South America:
Argentina 10,461
Brazil 103,262
Chile 1,863
Colombia 7,546
Venezuela 6,768

5

Africa: 576,972
Botswana 5,337
Burundi 8,378
Cameroon 8,141
Congo 10,223
Ghana 18,730
Ivory Coast 10,792
Kenya 69,005
Malawi 46,022
Rwanda 10,760
S. Africa 12,825
Tanzania 82,171
Uganda 51,779
Zaire 14,762
Zambia 42,447
Zimbabwe 61,037

6

Asia: 70,949
Japan 1,447
Thailand 59,782

7

Middle East: 3,000
Israel 433
Other 2,500

8

Oceania: 8,030
Australia 7,033
New Zealand 600

A total of 1,644,183 cases of AIDS have been reported to the World Health Organization as of January, 1997. A total of 208 countries are now reporting to WHO. Numbers by country are for various dates before January 1998.

Figure 2.2 AIDS cases reported to the World Health Organization (Not All Reporting Countries Are Listed).

C. AIDS CASES BY METROPOLITAN AREA, TO JULY 1999

Metropolitan area of residence (with 500,000 or more population)	Cumulative totals		
	Adults/ adolescents	*Children <13 years old*	*Total*
Akron, Ohio	518	1	519
Albany-Schenectady, N.Y.	1,561	24	1,585
Albuquerque, N.Mex.	1,005	2	1,007
Allentown, Pa.	752	8	760
Ann Arbor, Mich.	350	9	359
Atlanta, Ga.	14,757	103	14,860
Austin, Tex.	3,575	23	3,598
Bakersfield, Calif.	919	7	926
Baltimore, Md.	12,804	205	13,009
Baton Rouge, La.	1,666	19	1,685
Bergen-Passaic, N.J.	5,064	76	5,140
Birmingham, Ala.	1,722	22	1,744
Boston, Mass.	12,642	178	12,820
Buffalo, N.Y.	1,638	18	1,656
Charleston, S.C.	1,374	12	1,386
Charlotte, N.C.	1,902	22	1,924
Chicago, Ill.	19,128	219	19,347
Cincinnati, Ohio	1,780	14	1,794
Cleveland, Ohio	3,055	41	3,096
Columbia, S.C.	1,754	16	1,770
Columbus, Ohio	2,063	13	2,076
Dallas, Tex.	11,456	37	11,493
Dayton, Ohio	899	17	916
Denver, Colo.	5,256	19	5,275
Detroit, Mich.	6,998	73	7,071
El Paso, Tex.	962	10	972
Fort Lauderdale, Fla.	11,485	234	11,719
Fort Worth, Tex.	2,980	25	3,005
Fresno, Calif.	1,083	14	1,097
Gary, Ind.	648	3	651
Grand Rapids, Mich.	712	3	715
Greensboro, N.C.	1,474	20	1,494
Greenville, S.C.	1,335	4	1,339
Harrisburg, Pa.	910	7	917
Hartford, Conn.	3,661	46	3,707
Honolulu, Hawaii	1,667	12	1,679

Metropolitan area of residence (with 500,000 or more population)	Cumulative totals		
	Adults/ adolescents	Children <13 years old	Total
Houston, Tex.	18,001	153	18,154
Indianapolis, Ind.	2,650	14	2,664
Jacksonville, Fla.	4,008	68	4,076
Jersey City, N.J.	6,136	118	6,254
Kansas City, Mo.	3,662	13	3,675
Knoxville, Tenn.	657	6	663
Las Vegas, Nev.	3,223	26	3,249
Little Rock, Ark.	966	14	980
Los Angeles, Calif.	39,633	230	39,863
Louisville, Ky.	1,467	15	1,482
McAllen, Tex.	322	9	331
Memphis, Tenn.	2,704	15	2,719
Miami, Fla.	21,818	465	22,283
Middlesex, N.J.	2,946	69	3,015
Milwaukee, Wis.	1,791	16	1,807
Minneapolis-Saint Paul, Minn.	3,062	17	3,079
Mobile, Ala.	1,053	12	1,065
Monmouth Ocean, N.J.	2,632	61	2,693
Nashville, Tenn.	2,270	16	2,286
Nassau-Suffolk, N.Y.	6,155	109	6,264
New Haven, Conn.	5,973	121	6,094
New Orleans, La.	6,371	61	6,432
New York, N.Y.	109,899	1,971	111,870
Newark, N.J.	15,649	315	15,964
Norfolk, Va.	3,278	58	3,336
Oakland, Calif.	7,592	41	7,633
Oklahoma City, Okla.	1,550	8	1,558
Omaha, Nebr.	686	3	689
Orange County, Calif.	5,218	33	5,251
Orlando, Fla.	5,398	77	5,475
Philadelphia, Pa.	16,719	251	16,970
Phoenix, Ariz.	4,634	16	4,650
Pittsburgh, Pa.	2,208	17	2,225
Portland, Oreg.	3,597	8	3,605
Providence, R.I.	1,775	19	1,794
Raleigh-Durham, N.C.	1,778	21	1,799
Richmond, Va.	2,321	26	2,347
Riverside-San Bernardino, Calif.	6,404	51	6,455
Rochester, N.Y.	2,189	13	2,202

Metropolitan area of residence (with 500,000 or more population)	Cumulative totals		
	Adults/ adolescents	Children <13 years old	Total
Sacramento, Calif.	3,009	24	3,033
Saint Louis, Mo.	4,289	38	4,327
Salt Lake City, Utah	1,507	14	1,521
San Antonio, Tex.	3,671	28	3,699
San Diego, Calif.	9,876	52	9,928
San Francisco, Calif.	26,678	37	26,715
San Jose, Calif.	2,950	13	2,963
San Juan, P.R.	14,241	242	14,483
Sarasota, Fla.	1,275	21	1,296
Scranton, Pa.	404	4	408
Seattle, Wash.	6,262	19	6,281
Springfield, Mass.	1,486	24	1,510
Stockton, Calif.	686	13	699
Syracuse, N.Y.	1,163	10	1,173
Tacoma, Wash.	749	8	757
Tampa-Saint Petersburg, Fla.	7,642	98	7,740
Toledo, Ohio	529	10	539
Tucson, Ariz.	1,367	8	1,375
Tulsa, Okla.	1,015	8	1,023
Ventura, Calif.	753	3	756
Washington, D.C.	20,345	278	20,623
West Palm Beach, Fla.	6,692	195	6,887
Wichita, Kans.	653	2	655
Wilmington, Del.	1,785	15	1,800
Youngstown, Ohio	339	—	339

D. WHO ARE THE PEOPLE WITH AIDS?

Although it is clear that anyone engaging in known practices by which HIV is transmitted can contract the disease, 84 percent of AIDS cases in the United States are found in men. Over 70 percent of all cases are found in men between the ages of 25–49. Women account for 15 percent of all AIDS cases, up from 8 percent in 1987. Although young people (ages 0–20 years) account for 31 percent of the American population, only 2 percent of people with AIDS fall into this age range. Only .4 percent of people with AIDS are between the ages of 13–19 years. However, it is clear that a substantial number of people with AIDS who are now in their twenties became infected in their teens. Because of the long incubation period, AIDS simply does not appear until years after exposure to HIV. Thus, if a person is exposed to HIV in his or her teens, AIDS may not appear until

THE ECONOMIC IMPACT OF HIV/AIDS

a. *The person with AIDS.* The medical cost for people with AIDS from the time of diagnosis until death is now $102,000, up from $85,000 in 1991. Costs in terms of time lost from work will also be large because people with AIDS are in and out of the hospital and are increasingly unable to work as time passes. The costs multiply if anti-HIV drugs, such as AZT (see p. 52) are used. The new combination drugs prolong the life of persons with AIDS but this, in turn, increases the costs. THE FOLLOWING COSTS ALL INCREASED AT A FASTER RATE THAN IN THE PAST DUE TO THE BROADENED DEFINITION OF AIDS THAT WENT INTO EFFECT IN 1993 AND THE USE OF MULTIPLE DRUG TREATMENT (Currently $10,000 to $20,000 per year).

b. *Hospitals.* Almost all hospitals, especially those in metropolitan centers, face financial problems because: (1) they are obligated to protect their employees in all ways possible from contracting HIV, and (2) few persons with AIDS are insured enough or have enough savings to withstand many days in the hospital at costs that may range as high as several thousand dollars a day, depending on the level of care.

c. *Other disease research.* $1.4 billion was budgeted by the government for AIDS research in 1996. AIDS research now receives more government research money than any other disease in America. Some people worry that AIDS is drawing a disproportionately high share of research funds. For example, heart disease gets $1.2 billion for research but claims 15 times as many victims each year as AIDS. Cancer claims 10 times more victims each year.

d. *The general economy.* It is difficult to estimate the cost of AIDS to the general economy. There are direct costs such as research, hospitalization, and drugs, as well as indirect costs such as lost work time, loss of creative talent, and so forth. Current overall cost estimates range from $100 to $500 per year per American.

the person is in his or her twenties. Teenage sexual activity and drug use definitely contribute to the HIV infection rate. Do not be fooled into thinking you are safe from the disease as a teenager because so few young persons have AIDS.

AIDS is found in all races. In the United States, approximately 44 percent of people with AIDS are White, 37 percent are African-American, and 18 percent are Hispanic. In absolute numbers, the problem of AIDS is greatest in the White community. However, when you realize that African-Americans represent only 12 percent of the U.S. population, and Hispanics only 8 percent, it is clear that both groups are at high risk for contracting AIDS.

In the United States, **homosexual** (sexual orientation for those of the same sex) or **bisexual** (sexual orientation for either sex) males and **intravenous** (injecting drugs directly into the veins) drug users account for 80 percent of AIDS cases. Homosexual or bisexual men account for more than 83 percent of all Whites with AIDS, whereas they account for only 45 percent of African-Americans and 50 percent of Hispanic cases. About 42–44 percent of the AIDS cases among both African-Americans and Hispanics have been linked to drug abuse, while only 17 percent of cases among Whites have been so linked. Many heterosexual people who do not use drugs may feel that they have nothing to fear from AIDS. However, they are misinformed because people with AIDS have been found among all kinds of people and all age groups.

AIDS first hit the American male homosexual community. Responsible homosexual males quickly realized the danger AIDS presented to their lives and have been at the forefront of the fight against it. Once it was understood how HIV was transmitted, the gay community led the fight for changing behavior in the direction of "safer sex." Many homosexual men reduced the number of their sexual partners, avoided anal intercourse, avoided promiscuous sex as found in gay bathhouses and teahouses (public rest rooms used for anonymous sex), and used condoms to protect themselves. Unfortunately, recent research suggests that younger gay men are reverting to unsafe sexual practices. For example, 43 percent of 258 gay men aged 17–19 years surveyed in San Francisco reported practicing unprotected anal intercourse. Gay sex clubs seem to be increasing in popularity in some cities, although frequenting such clubs or gay bathhouses diminished early on in the AIDS epidemic.

In contrast to the United States, there are countries in central Africa where intravenous drug use and homosexuality are uncommon. In these countries, AIDS is epidemic equally among both women and men and is mostly spread by heterosexual relations. The African nations in question differ greatly from the United States. The different distribution of AIDS may be understandable in light of their different cultures and sexual behaviors. However, such countries teach us that AIDS can infect anyone, not just those persons who abuse drugs or engage in homosexual activity. The fact that only 10 percent of AIDS cases in the United States are classified as non-drug-abusing heterosexuals means that there is still a good chance AIDS can be prevented from spreading throughout the general population. It should be noted that of the adults with AIDS who are categorized as heterosexual (as of July, 1999 the figure was 70,582), most have had heterosexual contact with a person with AIDS or at high risk for AIDS (mainly sex with IV drug users, or were born in African or Caribbean nations with high rates of AIDS infection).

People with AIDS in the United States fall into the following categories (figure 2.3).

1. Sexually active homosexual and bisexual men (or men who have had sex with another man since 1977)   48 percent
2. Homosexual and bisexual men who are also IV drug users   6 percent
3. Present or past users of illegal intravenous drugs   26 percent
4. Persons with hemophilia or other blood clotting disorders who have received blood clotting factors   1 percent

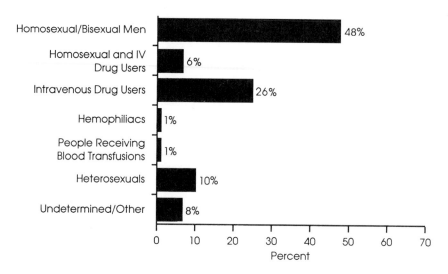

Figure 2.3   Distribution of AIDS in the United States

5. People who have had transfusions of blood, blood products, or tissues 1 percent
6. Heterosexual men and women (these include sex partners of persons
   with AIDS or at risk for AIDS) and people born in countries where
   heterosexual transmission is thought to be common   10 percent
7. Undetermined/Other   8 percent

As of January, 1999, more than 60 percent of all reported persons with AIDS have died (410,800). The typical patient dies within two years from the onset of symptoms. Presently the *mortality rate* (**the proportion of deaths**) of people with AIDS is over 90 percent. Thus, you can see why it is important for you to learn about this disease and how you can prevent it.

Although we hear mostly about AIDS, the virus that causes AIDS also causes other problems. Figure 2.4 shows that actual AIDS cases are really just the "tip of the iceberg." Another 600,000 persons are infected with HIV complex. (The CDC first called "HIV complex" AIDS-related complex [ARC].) These people have symptoms such as persistent generalized lymph gland (PGL) enlargement that persists for months and sometimes years. They suffer from fevers and night sweats and a general decline in health but have not yet developed the full-blown AIDS syndrome (see p. 12). About 7–10 percent of these people develop AIDS each year. HIV complex can be a fatal disease for some people; they can die from kidney disease, blood disorders, and other kinds of physical malfunctions.

The bottom of the iceberg, people who have been exposed to the virus, are carriers and can infect others even though they show no overt symptoms. The number of

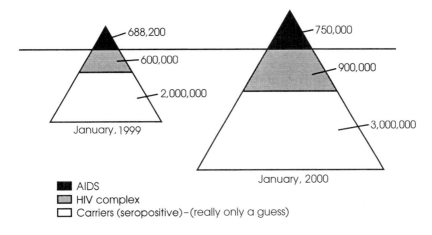

Figure 2.4   The Tip of the Iceberg

HIV+ people in America can only be a wild guess since testing of the general population for HIV has never been done. HIV/AIDS has become so politicized that to
suggest generalized HIV testing creates an immediate controversy about invasion of
privacy and confidentiality. Our best guesses about the number of HIV+ persons in
the population stem from the testing of Americans applying for military service.
Among the applicants, about 1 in 800, or .012 percent, are HIV+. Estimates suggest between 1 and 5 million Americans are HIV+. The other iceberg estimates
numbers for the three categories that are expected by January 2000. You can see from
these numbers that the HIV infection problem is actually much greater than the
number of people with AIDS officially recorded.

E. HOW HIV/AIDS AFFECT WOMEN AND CHILDREN

Currently 16 percent of people with AIDS in the United States are women. However,
the CDC is warning that women of childbearing age and children are becoming more
at risk. Although deaths from AIDS declined in men by 15 percent in 1996, deaths
increased among women by 3 percent. In 1985, women accounted for only 7 percent
of AIDS cases, while for 1996 the percentage of women diagnosed with the disease rose
to 20 percent. The cumulative number of U.S. women with AIDS reached 114,622 by
July, 1999. AIDS is the third leading cause of death among women aged 25 to 44.
Among teenagers, the proportion of female AIDS cases is about three times higher than
in adult females over 30. Lastly, looking at AIDS in third-world countries, one finds as
many, or more, infected women as there are infected men.

   The major risk factors for women are injecting drugs (43 percent) and sex with
an HIV-infected male partner (39 percent). Minority women and children appear to

be at higher risk. Of women with AIDS, 57 percent are African-American and 20 percent are Hispanic thereby accounting for 77 percent of all AIDS cases among women. The AIDS rate for African-American women is 17 times higher than for White women. AIDS ranks as the first cause of death among African-American women ages 25–44. The rate for Hispanic women is six times higher than for White women.

Although children under 13 years of age accounted for about 1 percent of all persons with AIDS, in 1998 there were 382 cases reported. Of these cases, 62 percent were African-American, 26 percent were Hispanic, and 16 percent were White or other. Nationwide it is estimated that one in 700 childbearing women is HIV+. In New York City, one in fifty childbearing African-American women is found to be HIV+. With women, it appears that poverty, poor education, and drug abuse are important co-factors to becoming infected with HIV.

Women are less able than men to protect themselves from HIV infection since they are often culturally and economically subordinate to men. For example, among intravenous drug users, women commonly use a drug apparatus after their male partners. Condoms are used less often because a woman must negotiate their use with her male sexual partner, who may feel reduced pleasure because of condom use.

During heterosexual intercourse women are more susceptible to HIV infection than men because semen with sperm and lymphocytes is often ejaculated directly into the uterine cervical canal, and semen normally contains 10–100 times more migratory lymphocytes than endocervical mucus.

Birth control pills make women more susceptible to HIV and other cell-borne sexually transmitted infections, because progesterones in the pills cause the uterine cervical opening to efface and, therefore expose more single layered epithelial cells through which lymphocytes can migrate easily (see fig. 3.3). (MOSTAD, Lancet Sept. 27, 1997, 922–927)

Among young women there is a close relationship between "survival sex" (sex used to obtain food, housing, drugs, and so forth) and HIV infection. For example, there is a strong association between the use of crack cocaine and the resurgence of sexually transmitted diseases, including HIV. Overall prevalence of HIV+ is 2.4 times higher in crack users than in noncrack users. Adolescent girls and young women, many of them not previously prostitutes, begin to traffic a large number of sexual contacts to support their crack habit. These young women seldom practice safe sex, such as condom use, in contrast to nonaddicted professional prostitutes.

The best short-term solution for women would be a safe and effective barrier to HIV under the control of women. The new female condom is one example. The use of cervical caps and diaphragms in combination with a microbicide are other examples (see part 4, section C, *If You Are Sexually Active,* figure 4.1, p. 45).

Women's progression from HIV+ to AIDS is about the same as men's, although the studies differ on time to death after AIDS diagnoses. Some studies report the time from first diagnosis of AIDS to death has been shorter for women than for men, but this is probably due to failure to recognize HIV infection at an early stage in women.

There are debates about ways to reduce heterosexual and infant transmission of HIV. Vasectomized men may be less likely to transmit AIDS, although there is not direct evidence of this. In Italy, a sperm washing technique has been experimentally successful in avoiding HIV transmission to infants when used in artificial insemination, because it removes most of the lymphocytes in the semen.

Children now constitute 1.4 percent of HIV disease cases in the United States. HIV appears to be transmitted relatively inefficiently from mother to offspring. The mechanism of transmission to the fetus or newborn appears to happen via provirus-infected lymphocytes that migrate from the blood of an HIV+ mother through the placenta at the time of labor and birth to infect her offspring. Current research suggests that there is a 25–35 percent fetal transmission rate. Interestingly, it has been found that women who give birth more than four hours after their "water breaks" (rupture of the fetal membranes, the protective sac that surrounds the fetus in the womb) are nearly twice as likely to transmit HIV to their infants, as compared to women who deliver within four hours of water breakage. The rate of transmission appears less for infants born via Caesarian section than born vaginally.

An HIV+ mother can infect her newborn infant by breast-feeding, although the exact risk is unknown. It is difficult separating those infants infected during pregnancy or birth from those infected through nursing. Most likely the transmission of HIV occurs in the early stages after birth when there is a high colostrum content in the breast milk. During this early stage, the mother's milk is characterized by high protein and antibody content. In industrialized nations where birthing and nursing conditions are relatively sterile, HIV/AIDS in infants can be avoided by using formula rather than breast milk. However, in third-world countries where conditions may be unsterile, newborns depend on colostrum, which provides maternal antibodies to protect the infant from many local diseases.

The incidence of perinatally acquired HIV/AIDS peaked in 1992, stabilized, and then began to drop. The CDC has recommended offering HIV testing to all pregnant women. There is a reduction in HIV transmission from mother to infant (25.5 percent to 8.3 percent) when both mother (in the predelivery stage) and infant are given continuing doses of AZT (p. 52). The dramatic decreases in pediatric AIDS clearly demonstrate the Public Health Services guidelines for AZT use during pregnancy. Counseling and testing of pregnant women have also been implimented and are decreasing the prenatal transmission of HIV (CDC *HIV/AIDS Surveillance Report,* December 1998, p. 7). Such a finding raises the issue of testing and prenatal screening. Should HIV-positive pregnant women be treated to reduce transmitting the infection to their infants? The other side of the coin is that untreated HIV+ pregnant women pass the virus on to only about 25–35 percent of their newborns. Thus, treating the other 65–75 percent of mothers-to-be with AZT raises ethical questions.

HOW DO YOU CONTRACT HIV/AIDS?

You read and hear about many different ways people contract HIV/AIDS. Some of the stories are true, but many are false. Unfortunately, you may be afraid of contracting HIV when you should not be. It is even more unfortunate if you are not afraid when you should be. Fear is not very helpful when you are trying to understand and make logical decisions. You need to understand and think clearly about HIV/AIDS. It is vital to your health to know the actual ways by which HIV is transmitted, as well as the ways by which it is *not* transmitted. This part of the book will address these issues.

A. HIGH-RISK BEHAVIORS

In the past, HIV has infected very specific groups of people in the United States (part 2, figure 2.3). By studying these groups and their behavior patterns, we have gained a better understanding of how HIV spreads and the kinds of precautions all people must take in order to avoid contracting or spreading the disease. Since anyone engaging in risky behaviors can transmit HIV, it is critical that we learn from the people who are infected.

High-risk groups' sexual behaviors and drug-related activities remain the major means by which HIV is transmitted in the United States. This information tells us that engaging in sexual activities or drug activities with multiple partners and without discrimination is dangerous. Primarily, HIV is spread through the sharing of virus-infected lymphocytes in *semen* (the thick, whitish fluid secreted by the male during ejaculation) and in blood. *Specifically, engaging in sexual behavior or injectable intravenous (IV) drug sharing with different partners (with little or no discrimination) are the major ways by which HIV is transmitted.* For example, if you have sex or share drug needles with two other people, and each of these people also shared needles or had sex with two other people and so on, you quickly find that you have interacted with literally hundreds of people. HIV also spreads through HIV-infected lymphocytes in mother's milk and occasionally through transfusion of blood, blood products, and tissues such as in an organ transplant. It can also be transmitted via artificial insemination.

You must understand that even if you personally do not engage in risky sexual behavior or use IV drugs, HIV can still be transmitted to you. It remains, essentially, but not exclusively, a disease of sharing sex, sharing drug-laden needles, and sharing blood.

B. SEXUAL BEHAVIOR

HIV is primarily a sexually transmitted disease (STD). Thus, the modes of transmission are much the same as with other STDs such as syphilis and gonorrhea. Throughout the world, sexual contact with persons who have the disease or carry HIV is the most common method of transmission. Avoidance of sexual contact (*abstinence*) is the most reliable method of preventing HIV infection. *Safer sex,* which you have been hearing so much about, involves avoiding sharing bodily fluids (semen in particular) (see table 3.1) by the use of condoms and other methods. However, so-called "safer sex" does not do away with the danger completely; avoiding sexual contact outside of a long-term, mutually *monogamous* (the habit of having only one mate) relationship is the only really safe sex.

Certain sexual practices are very likely to transmit HIV. Receptive anal-rectal intercourse (allowing the penis to enter one's rectum) appears to be the most dangerous sexual practice. This is true for both men and women. Because the lining of the rectum is a thin, single-cell layer, HIV-infected lymphocytes in semen can migrate into the body through breaks in the tissue (see figure 3.1). This area of the body is highly supplied with blood vessels, and insertion of the penis or other objects may result in tearing and bleeding. In addition, lymphocytes can migrate through the intact tissue under their own power, as shown in figure 3.1. *Anal-rectal intercourse must be avoided.*

Fellatio (kissing and insertion of the penis into the partner's mouth) and *cunnilingus* (kissing and insertion of the tongue into the vagina) may also be a means of HIV transmission. Saliva varies greatly in numbers of migrant lymphocytes (table 3.1) and, to date, has not played much of a part in the transmission of HIV. It appears to disrupt orally shed, infected leukocytes (*Archives of Internal Medicine,* Feb. 8, 1999, 303–311). Most authorities suggest that dry kissing is perfectly safe but that oral forms of sex may be unsafe. People with teeth or gum infections are particularly at risk, especially if ejac-

Table 3.1	
NUMBERS OF MIGRANT LYMPHOCYTES IN BODY SECRETIONS	
Circulating blood	4–20 million/teaspoon
Semen (pre-vasectomy)	±10 million/tsp.
Semen (post-vasectomy)	±1 million/tsp.
Maternal milk	4–12 million/tsp.
Endocervical mucus	±1,200 thousand/tsp.
Saliva*	±1,000 thousand/tsp.
Urine*	1–8 thousand/tsp.
Sweat*	Practically none found
Tears*	Small variable amounts

*Personal observations
Table from Dr. Jack W. Shields, Department of Medicine and Hematology, Santa Barbara Medical Foundation Clinic, Santa Barbara, California

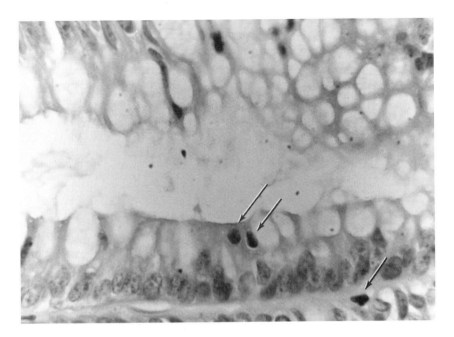

Figure 3.1   Rectum. Arrows show migrant lymphocytes moving through tissue. Credit: Dr. Jack Shields, M.D., and *Lymphology* 22:62–66 (1989).

ulation into the mouth takes place. The possibility that teeth may accidentally break the skin of the penis is another possible risk. Since the semen of infected males has a high concentration of HIV (table 3.1), it is probably not a good idea to swallow semen, although it is thought that acids and enzymes in the digestive system destroy HIV.

Vaginal intercourse can transmit HIV/AIDS to either men or women. About 10 percent of AIDS cases in the United States are reported from heterosexual relations and this percentage appears to be rising. The lining of the vagina is thick and difficult for the HIV-infected lymphocytes to penetrate (see fig. 3.2). In addition, the normally acidic (acid) environment of the vagina is not hospitable to lymphocytes, sperm, or HIV. However, the cervix and uterus have a single layer of cells that can be easily penetrated (see fig. 3.3). Semen is most likely to reach the uterus with prolonged or repeated acts of intercourse. Although, some authorities believe that cell-free HIV-1 can pass through multilayered vaginal mucosa (see fig. 3.3), cell-free virus remains to be demonstrated microscopically in genital secretions and lacks the power to move by itself. On the other hand, lymphocytes (such as the small one indicated by the arrow—lower left in fig. 3.3) customarily migrate back to regional lymph nodes via lymphatics which underlie surfaces.

Uncircumsized men are 10 times more likely to become infected than those circumsized. This is because the inside covering a redundant foreskin tends to remain damp, infected with a variety of bacteria and thin, especially in the region of the

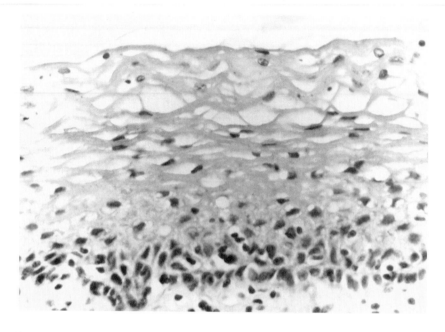

Figure 3.2   Stratified Epithelium (Vagina). No lymphocytes found migrating in superficial layers. Credit: Dr. Jack Shields, M.D.

frenulum which is relatively fragile. Thus, migratory lymphocytes in endocervical mucus secretions, although less numerous than lymphocytes in semen, are provided with relatively easy access to penile lymphatics.

Fortunately, for most men and women who enjoy regular vaginal intercourse, mutually monogamous sexual relationships, and avoid drugs, there is little danger of contracting HIV. The use of condoms, diaphragms, and cervical caps with a spermicide such as nonoxynol-9, attention to personal cleanliness, adequate lubrication, abstinence from sex during menstruation, and, of course, mutually monogamous relationships are critical to preventing the spread of HIV.

It should be noted that using condoms, engaging in sex only in a monogamous relationship, or abstinence are also the best methods of avoiding any sexually transmitted disease (STD). Other STDs, such as herpes, chlamydia, and syphilis, all increase the risk of AIDS. For example, women who suffer from genital ulcers (caused by syphilis, chancroid, and herpes) or chlamydia are two to ten times more likely to become infected with HIV. STDs cause lesions and breaks in the skin, thus allowing easier access to HIV-infected lymphocytes. Genital herpes appears to be a major factor in the sexual spread of HIV. University of Washington researchers have shown that people infected with HIV give off large amounts of the virus through herpes sores when they have flare-ups. About two-thirds of HIV+ persons also carry herpes virus type 2, the genital herpes virus which is also carried by lymphocytes.

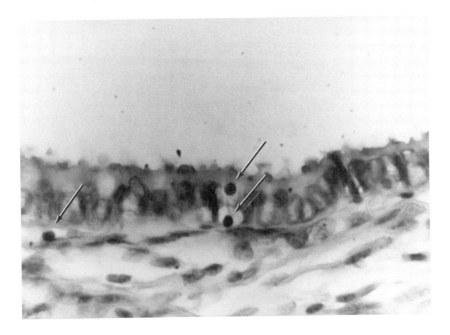

Figure 3.3   Endocervix (Uterus). Arrows show migrant lymphocytes. Credit: Dr. Jack Shields, M.D.

Another way STDs increase the risk of contracting HIV is immune system activation. Since HIV replicates within lymphocytes, any infection that activates a person's immune system not only produces more antibodies to fight the infection, but at the same time produces more HIV. This may account for why HIV/AIDS first hit gay men so hard. Hepatitis B and C, as well as other sexually transmitted diseases, were fairly widespread in the portion of the gay male population that included many sexual partners in their lifestyle.

Table 3.2 examines the effects of contraceptive barriers that help protect against HIV and other STDs, especially for women. Note: Other contraceptive methods such as oral contraceptives (the pill) or intrauterine devices (IUDs) give no protection against HIV/AIDS.

C. DRUG ABUSE

Intravenous (IV) drug users sharing syringes and needles are the second largest group of people with AIDS in the United States. They make up about 32 percent of people with AIDS. In some parts of the country, such as New York and New Jersey, IV drug users constitute the majority of infected people. IV drug users represent the

Table 3.2

METHODS OF SEXUALLY TRANSMITTED DISEASE PROTECTION, FROM WOMEN'S PERSPECTIVE

Method	Estimated Effectiveness Against Sexually Transmitted Disease	Contraceptive Effectiveness — High[a]	Contraceptive Effectiveness — Average[a]	Benefits	Disadvantages	Cost[b]
Condom	30–60%	98%	88%	Entails male responsibility; offers high level of protection; is inexpensive	Difficult to use properly; entails lack of control for women; may be seen as interrupting sex; may imply unfaithfulness	$0.50
Female condom	Insufficiently studied	Insufficiently studied	85%[c]	Offers high level of protection against sexually transmitted disease/HIV	Is visible; expensive; requires negotiation	$2.00
Film[d]	50%	99%	79%	Is easy to use; requires no negotiation	Requires 15 minutes' waiting time; must be applied within 1 hour of intercourse	$1.00
Suppository	50%	99%	79%	Is easy to use; is inexpensive; requires no negotiation	Requires 15 minutes' waiting time; must be applied within 1 hour of intercourse	$0.50
Foam	50%	99%	79%	Is available OTC[e]; requires no waiting time after insertion; requires no negotiation	Requires applicator	$0.50
Jelly/cream	50%	99%	79%	Is available OTC; inexpensive; requires no negotiation	Requires applicator; must be applied within 1 hour of intercourse	$5.00 per tube
Cervical cap	50–75% for cervical pathogens; 0% for others	98%	82%	Is comfortable; can be used repeatedly over 2+ days; is cheaper over reproductive life; requires no waiting time after insertion; no UTIs[g], rarely requires refitting; may require no negotiation (if not felt by partner); offers excellent cervical protection with low nonoxynol-9 use	Must be fitted; 20–40% women not able to be fitted; requires initial outlay of $100–$150; requires vaginal spermicide for best protection against sexually transmitted disease	$0.10

Method	Estimated Effectiveness Against Sexually Transmitted Disease	Contraceptive Effectiveness		Benefits	Disadvantages	Cost[b]
		High[a]	Average[a]			
Diaphragm	50–75% for cervical pathogens; trace (10%?) for others due to vaginal dispersion of spermicide	99%	82%	Requires no waiting time after insertion; fits nearly all women; may require no negotiation (if not felt by partner); is cheaper over reproductive life	Must be removed after 10–12 hours; may need to be refitted; carries increased risk of UTIs[g] for some women; requires initial outlay of $50–$75	$0.10
Withdrawal	Insufficiently studied	96%	82%	Requires no purchases	Is not controlled by woman; is highly user dependent	
Pill	None	99%	96%	Is convenient; is removed from sex act; affords high contraceptive efficacy	Offers no protection against sexually transmitted disease/HIV; may raise risk; is expensive	$12–$24/ per month
Norplant	None	99%	Insufficiently studied	Is convenient; is removed from sex act	Offers no protection against sexually transmitted disease/HIV; may cause bleeding in women, which may raise risk; carries high initial cost	$300– $500 per insertion; $5–$8 per month
Intrauterine device	None	99%	96%	Is not user dependent	Carries high initial cost; entails risk of PID associated with insertion	$150– $300 per insertion

Source: Golub, E. L., and Rosenberg, M. J. "Commentary: Methods Women Can Use That May Prevent Sexually Transmitted Disease Including HIV." *American Journal of Public Health*, November 1992, 1473.

[a]"Highest observed effectiveness; "typical" user effectiveness

[b]Per act of intercourse, based on average cost to consumer

[c]Use-effectiveness rate presented at Food and Drug Administration Hearings on Reality, January 31, 1992

[d]Marketed as "VCF" (vaginal contraceptive film) in United States and as "C-film" in United Kingdom

[e]Over-the-counter

[f]Rate depends on criteria for a good fit, and practitioner criteria have varied widely

[g]UTI = urinary tract infection

major bridge by which HIV spreads to women and children. This group serves as a bridge to the general population in three ways:

1. By sharing infected blood via needles used to inject drugs
2. By infecting a heterosexual partner via sexual contact
3. By infecting babies born of mothers who are infected via sexual contact or drugs

IV drug users are extremely hazardous because, unlike homosexual men who have organized to fight the spread of HIV/AIDS, they are engaged in illegal behavior and thus cannot openly organize to fight the disease. Many women IV drug users resort to prostitution to support their drug habits, and if infected, they add to the risk of spreading HIV via sexual relations. Also, when under the influence of drugs or alcohol, people behave differently. Thus, they may not practice lower-risk sexual behaviors and may use unsafe methods when *shooting up* (injecting drugs into the body with a needle and syringe). Sharing infected blood in needles is the major method of HIV transmission between IV drug users.

The use of crack cocaine is now recognized as another major influence increasing the risk of HIV/AIDS. The fact that sex is often traded for drugs increases the risk.

Unfortunately, drug users are difficult to reach. Attempts to educate this group on the dangers of HIV/AIDS and methods of prevention have not been very successful so far. Often the drug abuser does not care about his or her life nor the lives of others, since the drug habit already involves facing death from overdose, physical violence, and diseases. HIV/AIDS is simply one more hazard to face. As in the case with sexual transmission, *the best defense against HIV/AIDS is avoidance, simply not using drugs at all.*

D. BLOOD SUPPLY AND TRANSFUSION

More than 90 percent of the American population can be expected to receive a blood/blood products transfusion at least once in their life. Currently 12 million units of whole blood are donated annually to blood banks and another 15 million are bought from paid donors in plasma collection centers.

HIV transmitted by transfusion and the use of contaminated blood products accounts for only 2 percent (including the hemophilia category) of all AIDS cases in the United States. Hemophiliacs were particularly hard hit early in the epidemic because the anti-bleeding products necessary to their health are derived from many different sources. More than 4,000 hemophiliacs have already died from AIDS. The CDC estimates that 90 percent of those with severe hemophilia are now HIV+. A number of legal actions have been taken on their behalf against the blood product suppliers.

Fortunately, the chances of infection are decreasing since transmission through infected blood has been recognized. Of the 711,344 reported AIDS cases as of July, 1999, the CDC reported about 13,905 of those persons were infected by transfusion or the use of contaminated blood products.

HIV/AIDS TRANSMISSION VIA HOLLOW STEEL NEEDLES

*N*ext to sexual transmission of HIV, transmission by means of *hollow steel needles* is most important. When one realizes that more than 5 billion hollow bore needles are being used annually by 8 million health care workers and another half million dental care workers, and unknown numbers of persons using needles to shoot illegal drugs, it is understandable that possible needle misuse leading to the spread of HIV is a real worry.

The San Francisco Chronicle ran an extensive three part series on the dangers of disease transfer via needle sticks (April 13–15, 1998). It is estimated that there are about 1 million accidental needle sticks among medical and dental workers each year. Not only is there the threat of contracting HIV via a needle stick, but many other blood borne diseases such as Hepatitis B and C can also be transmitted. The series concluded that there is a virtual epidemic of blood borne diseases spread via accidental needle sticks.

New types of needle safety devices are now available. In one case a needle is covered by a sleeve which retracts as the needle enters the patient. As the needle is withdrawn the sleeve slides forward to cover the needle and locks in place. Unfortunately, only 5 to 10 percent of the syringes used by medical workers have the safety features ordered by the federal Occupational Safety and Health Administration, but never enforced. Such safety designed needles are more expensive, thus many hospitals and HMOs do not use them and, of course, those using needles for illegal purposes may not seek them out because of the increased cost.

The extreme hazards associated with medical use and illegal abuse of hollow bore needles comes into focus when one recalls that each teaspoon of blood normally contains approximately 10 million migrating lymphocytes or 50,000 to 150,000 per drop. Such a drop might contain HIV, Hepatitis B or C virus, herpes virus, etc. If it is then injected directly into the tissues or circulating blood without having to migrate through layers of cells lining skin, mouth, or vagina, it becomes the most efficient way to become infected.

Needle exchange programs for addicts have not worked well to date. First, many people object to such programs, feeling that they support, at public cost, the person's addiction. Second, legislation that would decontrol needle sales probably wouldn't work for the addict who is often short of funds. Third, there is a certain social comradery about sharing needles among drug addicts.

Since 1985, voluntary and paid blood donations have been tested for HIV contamination. The test is known as ***ELISA*** (enzyme-linked immunoabsorbent assay) and determines if antibodies against the HIV have formed. If the test is positive, the blood is discarded and the donor is notified confidentially of being HIV+. It does not mean that he or she has AIDS but only that there has been exposure to the virus. ELISA has helped to reduce HIV transmission via contaminated blood products. The tests are not perfect for a number of reasons, so the chance (although small) of HIV infection via contaminated blood or blood products still remains. The CDC suggests the chances of HIV infection via blood transfusion are about 1 in 420,000 (hepatitis B 1 in 100 and hepatitis C 1 in 200) units of blood, although some researchers suggest higher chances of infection. Unfortunately, antibodies to HIV infection do not appear for approximately six weeks to six months, thus leaving a "window of infection," i.e., an infected person will not be recognized during this time by the commonly used blood tests.

The Food and Drug Administration (FDA) has approved a test for HIV antigen screening of blood donors that will reduce the "window of infection" by a week or two. FDA still recommends the HIV antibody tests for routine testing. Home tests are also available as discussed earlier.

Most blood banks have established "look back" programs. These programs check on the source of blood given to recipients in the past and notify those recipients if the blood is discovered to be suspect. The blood recipient can then seek a blood test to see if there has been infection with HIV.

Although the blood tests are very accurate, there are still a few false positives and negatives. When a person tests HIV+ with ELISA, a second test—the Western blot—is then used to confirm the results. The CDC suggests that the two tests used together have a better than 99 percent overall accuracy rate. However, this still means that out of 10,000 people tested, 100 would receive erroneous results. Obviously, a false negative means an HIV+ person can infect others, all the while believing that he or she is not infected. A false positive can have a devastating effect on one's life. An uninfected person is led to believe that he or she is HIV+ and will at some time in the future develop AIDS. Hopefully, false results will continue to decline as testing experience increases. New tests for HIV are being developed.

E. CASUAL TRANSMISSION (THE TRANSMISSION OF HIV/AIDS WITHOUT SEXUAL CONTACT OR INTRAVENOUS DRUG USE)

The chances of casual transmission of HIV are very low. The surgeon general points out that HIV is transmitted in specific and limited ways, most often involving sexual contact or the sharing of needles among drug users. Outside of these specific ways of transmission, HIV does not seem to move from the infected to the noninfected person easily. If it were casually transmitted, it would be found scattered throughout the whole population by this time. Being near a person with AIDS is not considered dangerous unless you have sex with him or her or share needles and drugs. Although we cannot be 100 percent certain that casual transmission never happens, no instance has been clearly identified in the United States. So far, nearly all cases have been traced to the sharing of infected lymphocytes in semen, blood, blood products, and tissues.

Studies on medical personnel, such as the group at the San Francisco Hospital who have been taking care of AIDS patients since 1981, discovered no spreading of HIV through casual contact. Additionally, studies in large Eastern hospitals and in laboratories researching the AIDS viruses have shown no signs of AIDS or development of positive HIV tests among the medical and laboratory personnel (who are not in high-risk categories), except those who accidentally stuck themselves with blood-contaminated needles.

Recently the American Medical Association has reemphasized the necessity for health-care workers to be more consistent in *washing their hands*. Changing to a new pair of sanitized latex gloves every time one works with a new patient or outside of a sterile field is very important. The inherent capacity of latex to hold bacteria and other materials means the gloves may protect the wearer but pass on viruses, bacteria, and so forth to patients.

Multi-dose vials should be phased out. The latex covering on multi-dose vials through which the needle must penetrate can be contaminated unless cleaned carefully between uses. Even then there may be contamination. *Single dose vials* do away with this potential problem and thus should replace multi-dose vials as soon as possible.

In order to protect health-care workers from possible HIV infection, the CDC offers the following guidelines:

1. Prompt washing of hands or other skin surfaces following contact with bodily fluids and immediately after removal of gloves
2. Avoidance of injury with needles and other sharp instruments
3. Routine use of appropriate barriers such as gloves
4. Avoidance of direct patient care by health-care workers with skin lesions
5. Strict adherence to infection control by pregnant health-care workers to minimize risk of perinatal infection

It should be noted that healthy skin is a good barrier against HIV infection. However, chapped skin, skin rash, or any other kinds of diseases or break of the skin renders the person susceptible to HIV transmission through contact with bodily fluids.

Other evidence against the casual transmission of HIV indicates that, since 1983, when the viral causes of AIDS were first reported, the percentage of unexplained AIDS cases has steadily decreased. Cases that at first appeared to be examples of casual transmission have almost always been found to involve high-risk behaviors after thorough investigation. Family members caring for relatives with AIDS do not appear to contract the virus. As the CDC has suggested, no one is likely to contract HIV/AIDS by shaking hands, patting a friend on the shoulder, superficial kissing, drinking from the same cup, working in the same office, eating food served by infected persons, being sneezed or coughed on, using public rest rooms, swimming in a public pool, and so forth.

One story of HIV/AIDS transmission suggests that biting insects such as mosquitoes might spread the disease. There is no evidence at present to suggest insect involvement in the transmission of HIV/AIDS. Studies on insects reveal that they do not reproduce AIDS viruses, but there remains a remote possibility that they might carry small quantities of blood from one person to another. Whether such a small amount of infected blood could infect another person is highly doubtful. However, insects can carry other diseases, so it is a good idea to protect yourself.

The possibility of patients being infected by doctors and dentists who have HIV/AIDS was widely publicized in 1991, when a dentist with AIDS was found to have infected six office clients. After a thorough investigation, the CDC concluded

THE IMPACT OF HIV/AIDS ON HEALTH PROFESSIONALS

*I*f it is possible for infected medical personnel to infect patients with whom they work, then several questions must be raised. Does the patient have the right to know that the doctor or dentist is HIV+/AIDS infected? Do all medical personnel have an obligation to be tested periodically for HIV? Does the medical or dental profession or do the state licensing boards have the obligation to prohibit HIV/AIDS-infected personnel from working with patients?

Answers to such questions raise many ethical problems. Surely patients have a right to know whether their doctor or dentist is infected. On the other hand, to require all doctors and dentists to be tested and then prohibit those persons found to be infected from working, also raises many constitutional questions. The debate over infected medical professionals remains controversial.

The opposite side of this debate is equally important. Does the health-care worker have the right to know that a patient is HIV/AIDS infected? Obviously, if the patient has the right to know, so does the health-care worker. The possibility of subjecting innocent people to a fatal disease makes knowledge of HIV/AIDS infection important to both patients and health-care workers.

The CDC currently suggests that "invasive procedures" where there is likelihood of transmission of HIV to a patient be clearly identified by panels of specialists in different fields of health care, and that health-care workers known to be HIV+ voluntarily stop performing such procedures.

To be identified as HIV+ effectively ends the career of a health-care worker, while transmission of HIV to a patient subjects the latter to a fatal disease. Hopefully, clearer concepts and adherence to "universal blood and body fluids precautions," as well as honest exchanges of critical information before a needle or scalpel is inserted within health-care settings, will help to resolve this extremely difficult problem.

that all were infected from the dentist's blood during office procedures (each requiring the use of local anesthetics). Thus, in all probability, HIV was transferred via improperly sterilized needles.

Currently, in the United States there are approximately 500,000 practicing doctors and 150,000 dentists at risk for acquiring HIV in the workplace, or at risk for transmitting HIV-1 while at work, along with two or three assistants for each. Of all these health-care workers, approximately 18,856 have been reported with AIDS since the beginning of the pandemic. The CDC suggests that the occupational risk of HIV for all U.S. health-care workers lies somewhere in the neighborhood of 0.3 percent per accidental needle stick. The actual risk of HIV cross-infection between health-care workers and patients or between patients in health-care settings by means of needles, syringes, multiple dose vials, improper use of latex gloves, and inadequate handwashing during the use of the former is unknown, but is a risk that can be minimized by education.

4

HOW CAN YOU PREVENT AIDS?

You have read about the many physical complications that arise from HIV/AIDS, commonly resulting in sickness, disability, and death. We have considered how HIV/AIDS is spread among people. Since there are presently no safe preventive vaccines or totally effective medical treatments, it is clear that HIV/AIDS will continue to spread unless reasonable precautions are taken.

The future of the HIV/AIDS epidemic in the United States depends largely on each person. Each of us must assume responsibility for our own behavior, especially sexual and drug-related behavior. ***Whether or not HIV/AIDS becomes increasingly widespread is really up to all of us!***

The key to prevention lies in understanding that HIV/AIDS is primarily spread by the intimate sharing of bodily secretions. Sexual sharing of HIV-infected cells (lymphocytes) in semen is responsible for at least 54 percent of all AIDS cases in the United States. The sharing of blood by intravenous drug users is responsible for another 32 percent of all cases. Virus-infected mothers can spread the disease to their infants during pregnancy or birth by means of infected lymphocytes, which pass through the placenta or through the mother's milk during nursing. Saying "No" to drugs and avoiding indiscriminate sexual relations are the best preventive measures that can be currently taken against HIV/AIDS infection. The establishment of caring and meaningful intimate relationships will greatly reduce the spread of HIV/AIDS. ***Caring love*** is the real road to safer sex.

As noted earlier, taking precautions against HIV/AIDS also means precautions against many STDs. Twenty years ago, about five STDs were found and thought to be under control. Today there are many more STDs, such as herpes, hepatitis B, chlamydia, and HPV (human papilloma viruses, or genital warts). Other STDs are spreading even faster than HIV in some communities. Behaving in ways that reduce HIV risk also serves to reduce the risk of acquiring many of the older as well as newer STDs.

A. ESTABLISHING AND MAINTAINING CARING AND LOVING RELATIONSHIPS

Perhaps the most important thing to recognize is that HIV is spread by careless intimate relations. Those persons who have meaningful, caring, loving, intimate relationships with other persons, and who follow a few simple precautions, will not be at risk for HIV/AIDS, nor will they spread the disease. Unfortunately, the "sexual revolution," America's high divorce rate, and the generally free and accepting nature of our society make it harder to establish such meaningful intimate relationships. Historically in the United States, there were widely accepted rules regarding sexual behavior. Such rules made it relatively easy for the population to control sexual expression. Today the rules often tend to be individualized, with each person making up his or her own mind as to how to behave. It is especially difficult for young adults because they might not have established firm standards by which to make decisions concerning intimate relationships. Peer pressure may cause you to doubt your personal beliefs in the face of criticisms.

You might find it strange that we are discussing beliefs and lifestyles while considering HIV/AIDS prevention. However, your beliefs about the role of sex and drugs in your life will either help to spread HIV/AIDS or help to prevent it. Currently there is no medical cure for HIV/AIDS nor a medical means of prevention. As we will see later, it will be many years before an effective vaccine is developed. Thus, our best hope of containing HIV/AIDS is to change behaviors that spread the disease.

Since one's behavior is the most important factor in the spread of HIV/AIDS, it is strange that traditional public health procedures used in the past to control infectious diseases have been put aside. Persons putting aside such procedures have been partially responsible for the continued spread of HIV/AIDS. Such steps as partner tracking and notification, testing, and reporting have worked in the past to control behavior precipitated epidemics. For example, routine testing on all pregnant women for HIV would almost eliminate infants becoming HIV+. As noted, using AZT with pregnant HIV+ mothers, followed by the use of commercial milk rather than breast milk, would drastically limit maternal HIV transmission to infants. Routine HIV testing would give us some idea of the severity of the epidemic and the success rate of various attempts to control HIV spread.

If you have caring, meaningful relationships with people, then you will behave in responsible ways. People who care about others, care about themselves. As William Shakespeare's Hamlet said, "This above all: to thine own self be true, and it must follow, as the night [follows] the day, thou canst not then be false to any man." (Act 1, Sc. 3, 78–80). People who really care about themselves do not harm themselves or others. Responsible people voluntarily take the HIV test and share the results with their partner.

What are the qualities of human relationships that lead to warm, caring, happy, and meaningful relationships?

1. ***Commitment.*** When you are committed to another person or to your family, it means that you will stand by them and that they can count on you. You will not behave in a manner that is damaging to the relationship or to the other person. For example, if you are committed to your relationship and know the risk of HIV/AIDS if you have sex with different persons, then you will control your sexual behavior out of respect for yourself and your relationship with others. In other words, each partner can trust the other not

to engage in harmful behaviors. Such trust fosters care and love between the partners. Without trust, it is almost impossible to maintain a meaningful relationship of any kind.

2. *Appreciation.* Commitment and trust tend to help partners appreciate one another. You have probably felt bad when someone you love or a good friend failed to appreciate you or something good that you may have done. Sometimes you may treat poorly someone you love. If you bump into a stranger accidentally, you will probably excuse yourself. If you bump into your brother or sister, you may exclaim, "Look where you are going" or "You are so clumsy!" rather than apologizing even when you know it is your fault. If you get into the habit of treating those you care about with appreciation, focusing on their good deeds and actions rather than on their negative qualities, you will often find that the appreciation is returned. When you are appreciated by other people with whom you are close, you want to do good things for them. Thus, it is much easier to control behaviors that might expose you and your friends to HIV/AIDS.

3. *Good communication patterns.* If you can talk to your parents, to your brothers and sisters, to your friends, to your loved ones, to your teachers and school counselors about HIV/AIDS and the risky behaviors that spread the infection, you can learn how to avoid the inherent risks and help others to do the same. Because HIV/AIDS is mainly spread through sexual relations and through drug-related activities, it is especially hard to talk about it. It is hard enough to talk about sex with family members, much less the use of illegal drugs. Hence, keeping communication open and honest between yourself and others is important.

4. *A strong value system.* Most mature people have a value system by which they make decisions about life. Value systems are acquired. Whether it be formal religion such as Christianity or Judaism, or learning from those you most admire, it is clear that when *you know who you are* and *what you stand for,* choices are easier to make. If you develop a strong set of values, then it is easier to make your own decisions and avoid being unduly influenced by others. Developing and maintaining a strong value system is a lifelong process. As new situations and experiences arise, they must be integrated into your value system. HIV/AIDS is forcing some people to rethink and restructure their sexual values and their attitudes toward the use of drugs.

5. *Sexual and drug responsibility.* Sexuality is a strong force in everyone's life. The question is, how do you use sexuality in ways that will foster care and love both for you and for your partner? Certainly to risk HIV/AIDS infection does not indicate care and love. The same may be said for drug use. The evidence is clear that drugs harm you physically and psychologically. Thus, to use drugs is really a statement about your value system. It says that you really don't love or respect yourself, or care about others that are close to you. To behave in ways that risk HIV/AIDS is really a strong statement about your own lack of care for yourself. **An important step that each of us can take to avoid HIV/AIDS**

and thus stop the spread of this disease is to try and build a value system that will lead to caring and respectful relationships.

B. SPECIFIC PRECAUTIONS CARING PERSONS CAN TAKE TO AVOID HIV/AIDS RISK

There are a number of important steps that you can take to help yourself and others prevent the spread of HIV/AIDS. First of all, you can avoid both sexual relations and drug use, especially with multiple partners or people that you do not know well. *Abstinence* (avoidance) from both drugs and sex is still the safest way to prevent HIV/AIDS at the present time. Yet, in our free and easy society, it is especially hard to say "No" to sex. The advent of the birth control pill, bringing with it a revolution in sexual traditions, has made it as difficult to say "No" to sex as it used to be to say "Yes." The young person who says "No" to sex may be criticized by friends as being "uncool," "old-fashioned," "square," "a nerd," and so forth. Sexual feelings are strong and hard to resist. Lines such as "Everyone does it," "If it feels good, do it," "If you loved me you would," are hard to resist. Yet, to say "Yes" today puts one at risk of HIV infection as well as of other STDs. In addition, the woman takes the chance of becoming pregnant. Many young people do not necessarily deliberately plan to have sex with their date, but end up saying "It just happened." *Sex is too risky an activity to let it "just happen."* There are now life threatening consequences by not acting responsibly when it comes to sex.

Teenagers are physically mature enough to have sexual intercourse and become pregnant. However, emotions generally take a lot longer to mature to the point where two people can handle a sexual relationship and be good parents. For example, you hear a lot about child abuse. Studies indicate that much child abuse occurs with very young parents who did not want to be parents, and the abuse "just happened." Such parents are often not emotionally mature or experienced enough to cope with the frustrations involved in parenting and often react with anger.

We started this discussion by talking about the importance of meaningful relationships. Everyone wants to be cared for and loved. Everyone has sexual feelings. Yet, before we express these feelings we need to think about our own value system, our potential partner, and his/her feelings. We need to ask ourselves questions such as:

1. Will having sex with this person add to the relationship?
2. Am I (and is my partner) approaching sex freely? Am I being talked into having sex? Am I being made to feel guilty if I refuse sex?
3. Is having sex in agreement with my own value system? my partner's?
4. If I have a child, am I responsible enough to provide for it emotionally and financially?
5. If the relationship breaks up, will I be glad I had sex with this person?

6. Would those I care about approve of my having sex with this person?
7. Am I willing to risk a sexually transmitted disease, especially something as deadly as HIV/AIDS?
8. Can I handle the guilt and conflict I may feel?
9. Will my decision hurt others? my partner? my friends? my parents?
10. Do I know how to protect myself, both from sexually transmitted disease and pregnancy?

These questions are not easy to answer, especially when you feel a strong sexual attraction toward another person. This is why you need to think about your answers to such questions before you find yourself in the position of having sex. To be parked in a car in a romantic spot or actually in bed with your lover is not the time to try and answer such questions.

If you need time to think about such questions and yet find yourself in the position of having to make an immediate decision, it is best to say "No." This can be said in such a way as to still remain friends and not hurt the other person's feelings. For example, you can say things like, "I like you a lot, but I'm just not ready to have sex." "I enjoy being with you, but I don't think I want to have sex." "I don't believe in having sex before marriage. I want to wait." "I don't feel like I have to give you a reason for not having sex. It's my decision." In light of HIV/AIDS, it may be that saying "No" to sex and drugs is the best way to say "I Love You" and "I Love Me."

C. IF YOU ARE SEXUALLY ACTIVE

Sexual intercourse just for fun, supposedly without attachments, regrets, or responsibility, sounds like a happy concept. However, the possibility of an unwanted pregnancy and/or a STD really means that sexual relations must always be considered carefully and responsibly.

1. It is important to limit your sexual activities. The more partners you have, the greater the risk of HIV/AIDS as well as other STDs. Monogamous relationships can be meaningful even without sexual intercourse. Talking, touching, holding hands, kissing, fondling, cuddling, snuggling, and just enjoying one another's company can all be exciting and stimulating and can lead to meaningful relationships.
2. It is important that you have respect for your partner and can trust that he or she is not engaging in unsafe sex practices. Researchers, such as Masters and Johnson, suggest that each partner be checked for HIV before commencing a sexual relationship. Some states have *voluntary partner notification programs* at test sites. Such programs offer HIV+ persons the option of having a trained health department counselor notify his or her exposed sexual or needle sharing partner(s) of their exposure to a person known to be infected with HIV. The identity of the HIV+ person is usually not shared and the counselor refers the partner for HIV testing and counseling.

THE IMPACT OF HIV/AIDS ON CONFIDENTIALITY

*C*onfidentiality versus the duty to warn is being faced by more and more health-care professionals as the HIV/AIDS epidemic spreads. A number of states have enacted laws that expressly forbid disclosure to anyone of an individual's HIV status. These laws have been passed because of the risks of discrimination that accompany AIDS. Yet, opposed to this is the health-care professional's "duty to warn." Legal precedent clearly points to protecting others in certain circumstances. Public health laws have often required the reporting and contact tracing of people with a variety of contagious diseases. An infected person's reckless behavior can pose a serious threat to the general society and needs to be controlled. Health-care professionals are finding themselves increasingly caught between these two opposing doctrines. In hindsight, it is clear that had traditional public health policies been followed at the outset, the HIV/AIDS epidemic would have been better controlled.

---

3. If you have questions or need additional information to make the impor-tant decisions about your sexual behavior, consult your community public health department, family physician, school counselor, and other adults in a position to help.
4. Personal cleanliness is important for all sexual partners. For males, this means regular bathing and bathing of the penis with special attention to the foreskin, especially for the uncircumcised male, as soon as possible after intercourse. Careful cleansing of the female genitalia can prove equally important, not only for the prevention of HIV infection, but also for the prevention of herpes and many other sexually transmitted diseases. However, no matter how clean part-ners are, sex outside of a monogamous, mutually caring relationship and with-out the use of a condom should be avoided.
5. Avoidance of sharing semen through the proper use of a condom is critical. The ***condom*** or ***rubber*** (see figure 4.1), is a sheath of very thin latex that fits over the penis and stops semen from entering the vagina when ejacula-tion occurs. Much advertising about the benefit of condom use in avoiding HIV makes it appear that condom use is the answer to the epidemic. Unfortunately, the condom is not perfect and must be used properly if it is to be effective. **Both you and your partner must realize the shortcomings of condoms.**

- It must be in place before sexual intercourse begins.
- There must be good lubrication so that it does not break.
- Some lubricants are petroleum-based such as Vaseline and deteriorate latex. They should be avoided in favor of water-based lubricants. Some condoms are prelubricated by the manufacturer. If not, K-Y Jelly works well as a lubricant.
- Sexual intercourse must cease once there is ejaculation, so that there is no danger of semen leakage.
- You must be careful not to puncture the condom.

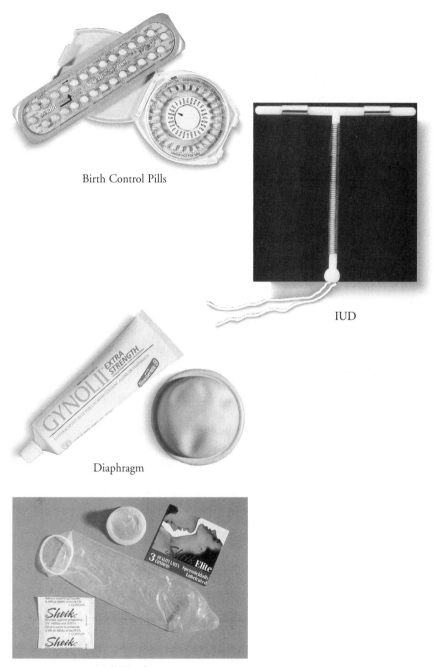

Birth Control Pills

IUD

Diaphragm

Male Condom

Figure 4.1   Various Birth Control Devices
© McGraw-Hill Companies.

- You must use a latex condom rather than one made of animal membranes since it has been found that the HIV-infected lymphocytes can pass through such membranes.
- The condom package must state clearly that the condoms are able to prevent disease. If the package says nothing about disease prevention, seek another brand.
- Condoms of American manufacture appear to have lower failure rates than those of foreign origin. It should be noted that condoms occasionally fail (approximately a 10 percent failure rate) due to manufacturing defects but more often failure is due to improper use. Thus condoms are not a foolproof guarantee of avoiding HIV/AIDS infection.
- Be sure that the condom covers the entire penis and that it is carefully removed immediately after ejaculation.
- Leaving a condom in a wallet, purse, or a warm place like a glove box for a long time will cause deterioration. If a condom sticks to itself, or is gummy or brittle, do not use it.
- Most condoms are generally not designed for and are prone to break during anal intercourse.

  Despite all of the efforts to promote condom use for protection against HIV/AIDS and other STDs, studies indicate that few people use condoms on a consistent and regular basis. Studies of condom usage vary, but most research indicates that only 25 to 35 percent of young persons engaging in sexual intercourse use condoms consistently. When one partner is HIV+ or has AIDS but the couple use a condom correctly and consistently the transfer of HIV to the uninfected partner appears unlikely. The safest method of AIDS prevention remains abstinence.

6. The *vaginal diaphragm* and **cervical cap** are contraceptive barriers specifically designed to prevent the access of semen into the canal inside the uterine cervix which leads on to the inside of the uterus. For centuries cervical caps were really the only effective women's choice for preventing pregnancy. The cervical caps or diaphragms became just as, if not more, efficient than condoms for this purpose, when low concentrations of non-oxynol-9 (N-9) were added to impair the motility of sperm. Their usage declined with the common use of birth control pills. They are still used by many discerning women.

As women's choices, the diaphragm and cervical cap have not been officially recommended for HIV/AIDS prevention, because they do not cover the vagina. However, in large trials they have proved more effective than condoms in prevention of cell-borne sexually transmitted diseases, such as gonorrhea, chlamydia and hepatitis infections. Their use as an effective women's choice for contracepting AIDS might follow when it is generally recognized that (a) HIV is lymphocyte-borne; (b) lymphocytes do not normally migrate through multilayered vaginal cells; (c) birth control pills invite AIDS by effacing the uterine cervix to expose single layered cells; (d) lymphocytes are 10–100 percent more sensitive than sperm to the detergent effects of N-9 and (e) they are available and reusable for months. There are a number of advantages for diaphragms and cervical caps:

- They are reusable, making them relatively inexpensive in the long run.
- They do not interfere with male or female vaginal sensation.
- They permit ejaculation within the vagina without immediate withdrawal of the penis.
- They can be put in place well before sexual contact and left in place for 48 hours after intercourse, thus not interfering with the act of loving.
- Once in place, they are not uncomfortable and do not interfere with other activities.
- They give the woman control.

Disadvantages are:

- Vaginal diaphragms and cervical caps cost about $30 apiece, must be properly fitted by experienced technicians, and require a course of initial instruction for the user.
- They require cleansing and care because they are used over and over for six to eighteen months.
- They should not be worn during menstruation.
- Models made from latex should not be left in the sunlight or used with petroleum-based lubricants.
- They may cause uterine cervical or vaginal irritation, especially if left in for longer periods of time.
- They are a nuisance, requiring skills to put in and take out properly and therefore used mostly by women accustomed to planning ahead.

The female condom is a single use contraceptive device. Unfortunately it costs about $2, thus making it out of reach financially for some people. It consists of a polyethylene tube, one end of which encloses the female cervix by means of an inner ring and an outer ring holds the other end at the entrance to the vagina. During intercourse the penis enters the tube and does not have direct contact with the vagina. It is somewhat awkward to use. Other female condoms are being developed. The major advantage over the male condom is that it gives the woman the ability to protect herself and her partner. She does not need to rely on her male partner for protection during intercourse. Of course, such contraceptive devices are designed primarily to prevent pregnancy. However, if pregnancy does occur, the prospective parents or at least the mother should seek public health guidance as soon as possible to find out if an HIV infection is involved.

The CDC is researching topical microbicides which could be in the form of gel, foam, or suppository for women to use to kill HIV before it has a chance to enter the lymph system.

- Consider that your chances of getting HIV/AIDS are greatest in sexual relationships with older persons, as well as homosexual or bisexual men and persons shooting up drugs.

- Avoid anal intercourse, because the thin lining of the rectum makes this the most hazardous kind of sex for the transmission of HIV/AIDS.

D. DRUG USE

1. **Total avoidance of intravenous drug use is crucial to preventing HIV/AIDS.** Sharing needles risks transmitting the virus-infected lymphocytes in the blood, and this is the most direct and efficient way to become infected. Saying "No" to drugs means saying "No" to HIV/AIDS.
2. When you are under the influence of drugs or alcohol, your judgment is often poor. Poor judgment leads to accidents, injuries to yourself and your friends, and HIV/AIDS, if careless sharing of sex and/or drugs via needles is involved.
3. You must remember that drug addicts are driven to do anything necessary to get a drug fix. When the addict is in need of a fix, concerns such as HIV/AIDS prevention will not be considered. The desire for drugs simply blocks out all other concerns.
4. Although controversial, some people advocate needle exchange programs. In most cities and states, sale or purchase of hypodermic needles and syringes without a prescription from a medical practitioner are prohibited. The idea behind needle exchange is that needles will not have to be shared and reused (increasing the chances of infection) if new ones are easily available to addicted people. Several cities such as Seattle, Washington; Boulder, Colorado; and New York City, have started needle exchange programs. Several studies reported in the *Journal of the American Medical Association* (January 1994) suggest these needle exchange programs have reduced needle sharing. The studies also found no evidence that such programs induce more people to shoot up. However, it remains unclear as to just how successful such programs are in reducing the spread of HIV. The use of bleach to clean needles is suggested but there is little or no evidence that this kills HIV.

E. DONATE TO BLOOD BANKS

There is absolutely no way to contract HIV/AIDS through donating blood. Most blood collection agencies such as the Red Cross require you to be 17 or older. You can, however, donate your own blood for yourself if you are going to need it in the future (for example, if you have to undergo surgery and might need a blood transfusion). Transfusions of blood or blood products are important to the health of the nation. In fact, 90-plus percent of Americans can expect to have one or more transfusions of blood or blood products in their lifetime. It is critical that our blood supplies be kept free from all disease-producing viruses.

The key to prevention of transfusional HIV/AIDS is to maximize the number of safe donors and minimize the numbers of high-risk donors. Thus, as a person who remains free of HIV/AIDS and other blood transmitted diseases, you should plan to give blood periodically. Even if you are too young at this time, you can still encourage your older, healthy friends and family members to donate, and plan to donate when you are old enough.

Lives will be saved if you and your friends understand that the spread of HIV/AIDS can be stopped. But it can only be stopped if everyone works together to change high-risk behaviors that spread the disease. Understanding HIV/AIDS and how to stop its spread is crucial to your future health as well as that of the nation.

F. VACCINES, DRUGS, AND CURES

In the recent past, modern medicine has had great success treating many diseases, including STDs. Many people do not worry about contracting an STD because they think that a shot of penicillin will quickly cure them. Unfortunately, with AIDS no cure exists. The Tenth (August 1994) and Eleventh (July 1996) International Conferences on HIV/AIDS found that science, medicine, and governments have made some progress in the treatment of HIV/AIDS, but the conferences clearly made the point that education and prevention remain our best weapons in the fight against HIV/AIDS. Progress has been so slow that international conferences are now planned for every other year rather than every year as in the past.

Vaccines are imposters, harmless foes intended to be perceived by the body as enemies. When the deception works, the harmless imitators offer the immune system a crash course in self-defense that is remembered for years and serves to stop the real disease if it is encountered. Currently, about three dozen preventative HIV vaccines are being tested in small-scale clinical trials around the world. Since 1988, more than 2000 healthy non-HIV-infected adults have voluntarily enrolled in 25 Phase I and II experimental vaccine trials conducted in the United States.

One problem with HIV vaccines is the discovery of other viruses similar to the original AIDS virus. In addition, just as there are constantly changing flu viruses (mutations) requiring flu vaccine modifications, there are mutations of the HIV. If mutations of the HIV occur, a vaccine that effectively stops one strain of the virus from reproducing might be ineffective in stopping another strain of HIV.

Despite serious efforts to produce a viable vaccine, there is nothing currently on the horizon. At best, a workable vaccine is probably some years away and some scientists even suggest that it is impossible to produce a vaccine against HIV. A number of vaccines have already been tested for safety. However, large-scale human testing has been postponed because of lack of promise of the various vaccines. As noted earlier, some critics suggest that the large amounts of money spent on vaccine research would be better spent reinstating the traditional public health rules for fighting epidemics, because it is one's own behavior that leads to HIV infection. By curbing HIV/AIDS risk behavior, the need for a vaccine becomes moot. Practicing HIV/AIDS risk behavior in the hope that a cure or vaccine will save you is suicidal.

Drug Therapies   You may have read the exciting news about the use of multiple drugs (so-called "drug cocktails"). These have been helpful in slowing the progress of HIV/AIDS for some infected people. The key word is "help" rather than "cure." Medicine is able to fight some of the many of the opportunistic illnesses that attack a person with AIDS, such help serves only to postpone the inevitable death. For example, *AZT* (*azidothymidine,* trade name Retrovir, and the first drug to receive FDA approval for fighting AIDS) does act to reduce AIDS symptoms and to prolong the life of people with AIDS. Unfortunately, its usefulness wears off over time. Studies indicate that, at best, drug interventions appear to prolong the life of a person with AIDS for several years. To improve one's quality of life and increase the number of years that a person with AIDS can live comfortably is a worthy goal but it is not a cure. In addition, there are strong negative side effects for many patients on this drug. These side effects include anemia, requiring frequent blood transfusions in up to 40 percent of patients and reduced white blood cell formation in 20 percent of the patients. This necessitates stopping the use of the drug for fear of other infections. AZT induces a rash, itching, headaches, and mental confusion in some patients.

AZT treatment of asymptomatic HIV+ people has been thought to slow the onset of AIDS. Studies reported at the Ninth International Conference on AIDS found there are individual reactions to AZT and, for a few persons treatment with it may, in fact, speed the onset of AIDS. Because of the side effects, giving AZT to an asymptomatic patient feeling well creates a person who feels sick.

Many new drugs have been introduced since AZT was approved (see appendix C). The newer protease inhibitor drugs given in combinations (usually of three drugs, "triple combination therapy") have excited researchers who find that AIDS development in HIV+ people is delayed.

The media excitement about these new drugs tended to drown out the researchers' own cautions about the success rate of the drugs. HIV/AIDS researchers' meeting in June of 1997 pointed out that the triple combination therapy can fail for a variety of reasons. First, complying with the therapy requires taking dozens of pills each day, many of which have serious side effects and dietary restrictions. If the regimen is not followed carefully, drug-resistant mutations are given a chance to appear. As one researcher said, "A little bit of treatment won't work. Once you've decided to treat, you must be aggressive." But starting this complex drug treatment locks patients into an expensive program of doctor visits, the threat of debilitating side effects, and an inflexible lifestyle built around taking the next pill. The combination-of-drugs treatment calls for two to three drug sessions daily, involving 15 to 20 pills. Second, since there are numerous drugs currently available, each with a slightly different chemical structure, the reaction of patients to a given drug can vary widely. It is important that a patient be given the proper combination for him/her. New federal HIV/AIDS treatment guidelines to help doctors determine the correct combination were released in mid-1997. At that time, there were eleven approved antiviral drugs that could be taken in any of 320 different combinations.

Another major problem with triple combination therapy is high cost. The pills have a wholesale cost of $10,000 to $20,000 a year. Obviously such a cost cannot be borne by most Americans, much less people in underdeveloped countries. In America, the AIDS Drug Assistance Program (ADAP) helps fund drug treatment for low income HIV+ persons as well as those with AIDS. The program is funded by combined state and federal contributions with the states contributing one dollar for every two federal dollars received. For 1997, the federal ADAP budget was almost $200 million.

Last, but most important, is the failure of the triple combination drug treatment to maintain reduced HIV levels after six months for half of those treated. It appears that about half of those treated by these new drugs will experience continued suppression of HIV while the other half will exhibit disease progression after time. Apparently those who evidenced failure after at least six months of drug therapy were people with very advanced HIV disease. Most had tried various drugs earlier in the course of their HIV sickness, which caused resistance to even the combination drugs. While clinical trials of combination drug therapies with selected patients showed an 80 to 90 percent success rate, the more typical patient population may have only a 50-50 chance of lasting benefit.

The drop in AIDS deaths commencing in 1996 is probably accounted for by the use of the new combined drugs therapy as is the drop in reported AIDS cases since 1996. Whether these are really meaningful signs that the epidemic is abating, or only a temporary artifact of the new improved treatments, (remember these new treatments do not appear to cure HIV/AIDS but rather keep it in check and prolong better health and life) remains to be seen. Also these rate drops mask the fact that cases among women and heterosexuals who do not use drugs are on the rise.

The FDA has speeded the process by which new drugs may be made available for desperately ill patients before being approved for full marketing. In addition, rules for importing non-approved drugs have eased. An individual may import a foreign drug for personal use only, but a licensed physician must prescribe the drug.

The business of HIV/AIDS has grown rapidly, since there is no quick cure for HIV/AIDS or a vaccine to prevent it. Companies realize that the disease will consume a growing share of the health-care budget in the future. The rapid commercialization and politicalization of HIV/AIDS has led to many speculative treatments and products. Care and caution must be used when one reads or hears media reports on some new surefire cure for HIV/AIDS.

At least for the near future, the only safe approach to HIV/AIDS is a caring, preventative approach. Presently, this is best achieved by learning how to avoid spreading HIV from one person to another. EDUCATION and PREVENTION remain our only tools for fighting AIDS. The building of loving, caring, respectful relationships is the foundation for fighting HIV/AIDS as well as for building a better society. People who care about others and themselves do not exhibit harmful behavior. People who care do not engage in behaviors that risk the transmission of HIV/AIDS.

THE IMPACT OF HIV/AIDS ON THE LAW

*A*IDS is having a wide and varied effect on America's laws and legal system. Many laws have already been passed protecting the rights of the person with AIDS. Anti-discrimination laws in the workplace and in housing have been passed. Laws allowing medical workers to be told if their patients have tested positive for HIV are in place. Insurance companies can require HIV/AIDS tests for new customers and refuse policies to those infected with the disease. One person assaulted by another person can obtain a court order forcing the accused attacker to submit to HIV testing. All of the laws stimulated by AIDS are simply too numerous too list.

It is important to note, however, that AIDS-related criminal laws and prosecutions are becoming more frequent. When people who are HIV/AIDS-infected knowingly engage in sex or needle sharing without informing their partner, they can be prosecuted criminally in some jurisdictions for "assault with a deadly weapon." It is possible that if a person died from AIDS after being knowingly infected by a sexual or drug partner, the partner could be charged with manslaughter. In addition, there is an increasing number of civil suits being brought against persons who do not inform their sexual and drug partners of their HIV+/AIDS-infected condition.

The following are just a few examples of how the law is becoming increasingly involved with HIV/AIDS:

- In October, 1991, Alberto Gonzalez (Oregon) became the first person in the nation to be convicted on assault charges for knowingly passing HIV to his girlfriend.
- In September, 1992, a Denver jury orders United Blood Services, of Albuquerque, New Mexico, to pay $6.6 million to a woman who contracted HIV from blood donated by an HIV+ homosexual male donor.
- In July, 1992, a New York state court orders the state to pay a Utica nurse $5.4 million after she contracted HIV from a prison inmate during a scuffle at a hospital.
- In July, 1992, a bill is introduced before the New Jersey state legislature that would make it a capital offense to cause death by knowingly infecting another person with HIV.
- Also in 1992, a surgical technician accidentally cut by a scalpel during a postoperative procedure was awarded $102,000. The damages for fraud and intentional infliction of emotional stress were assessed against a breast surgery patient who failed to disclose that she was HIV+. (See *The AIDS Litigation Project III: A Look At HIV/AIDS in the Courts of the 1990s* by L. Gostin funded and published by the Henry J. Kaiser Family Foundation July 24, 1996, for a complete report on the many legal decisions already taken concerning HIV/AIDS.)
- In 1998 the Supreme Court debated whether HIV+ people should be considered disabled and thus come under the Americans with Disabilities Act. A dentist refused to treat an HIV+ patient in his office but would treat the patient in a hospital setting. Is this discrimination?

SUMMARY: PREVENTING HIV/AIDS

The following precautions will help prevent you from contracting or spreading HIV/AIDS and other sexually transmitted diseases (STDs):

1. Sexual abstinence, especially when a caring relationship is not involved.
2. Sexual fidelity.
3. If you want to have sexual relations but are not in a permanent monogamous relationship, use mechanical barriers to prevent the exchange of potentially infectious body fluids, especially blood, semen, and uterine secretions.
4. Use barriers proven to prevent pregnancy as well as sexually transmitted diseases (STDs), such as:
   a. FDA-approved **latex** condoms that can protect a woman and a man from sharing semen and vaginal secretions during conventional vaginal intercourse.
   b. Doctor-prescribed and fitted vaginal diaphragms or cervical caps that block semen from reaching the uterus.
   c. In addition, use spermicides such as nonoxynol-9, which paralyze sperm and migrant lymphocytes that may have gotten past the barrier. A few people will have allergic reactions to specific spermicides.
5. Use barriers strictly in accordance with recommendations supplied by the FDA, as well as inserts supplied by the manufacturers. A great deal of careful research, time, and effort are embodied in these recommendations.
6. Avoid anal intercourse with or without a condom, because this is the most dangerous way to share semen; and condoms are not well designed for this sexual expression.
7. Avoid sexual relations with persons at great risk for being HIV+ or having AIDS or other transmissible viruses, such as homosexual or bisexual persons, or persons who "shoot" drugs or persons who sell or buy sex.
8. Do not use alcohol or drugs. They interfere with you caring for yourself as well as for others, and especially avoid the use of drugs that are injected into the veins.
9. Do not share needles used for injecting drugs into the veins or handle sharp instruments contaminated with another person's blood.
10. Donate blood to your community blood banks and encourage your healthy friends to do so as often as possible. Such donations will help save lives and protect the blood supply from possible contamination.
11. If you know well ahead of time that you might need blood during an operation, predonate yourself, and let your caring friends of similar blood type know that their donations would be appreciated.
12. Continue to build caring, meaningful, and trusting relationships with people.
13. Regarding persons with whom you are not intimate, who are sick with any disease, or are known to have AIDS, your sincere caring will prove to be mutually helpful.

14. Although AIDS seems destined to become one of the most deadly epidemics humans have faced, do not be afraid. Your chances of becoming infected are near zero if you apply the foregoing precautions.

15. Do not pay attention to rumors and hearsay about AIDS. Whenever you have a questions about AIDS or behavior that increases the chances of exposure, seek out answers from health professionals, medical clinics, teachers, counselors, and other knowledgeable adults.

16. You should not be lulled into a sense of complacency or apathy encouraged by the conflicting reports or lack of reports in the news media. Although significant "breakthroughs" are often reported, you must do all you can to protect yourself from acquiring HIV since all real and permanent cures still remain in the future.

17. When you seek diagnosis or treatment in a health-care facility, you must be aware that you and all employees are at increased risk for infectious diseases. Therefore you would be wise to protect yourself, along with your health-care providers, by insisting on adequate hand washing, proper use of protective gloves, better protection for syringes and needles, and no reuse of multiple-dose vials for giving injectable medications. Although your insistence might seem improper, it will lead to improved care and help to reduce infectious disease spread.

18. Finally, your health and that of your community is our greatest national asset. Let's not lose it by being careless.

THE CENTERS FOR DISEASE CONTROL'S REVISED DEFINITION OF AIDS AS OF JANUARY 1993

EXPANDED U.S. AIDS SURVEILLANCE CASE DEFINITION

- All HIV-infected adolescents and adults with fewer than 200 CD^4+ T-lymphocytes/μL
- Or HIV positive and one or more of the following:
  Cryptosporidiosis, cytomegalovirus, isosporiasis, Kaposi's sarcoma, lymphoma, lymphoid pneumonia (hyperplasia), *Pneumocystis carinii* pneumonia, progressive multifocal leukoencephalopathy, toxoplasmosis, candidiasis, coccidioidomycosis, cryptococcosis, herpes simplex virus, histoplasmosis, extrapulmonary tuberculosis, other mycobacteriosis, salmonellosis, other bacterial infections, HIV encephalopathy (dementia), HIV wasting syndrome, pulmonary tuberculosis, recurrent pneumonia, and invasive cervical cancer

Source: Centers for Disease Control and Prevention.

TEN POINTS FOR WORLD AIDS DAY: AIDS AND THE FAMILY

1. **HIV and AIDS.** AIDS (acquired immunodeficiency syndrome) is the late stage of infection with the human immunodeficiency virus (HIV). AIDS can take more than 10 years to develop, and most people die within 3 years of it being diagnosed.
2. **Modes of transmission.** The vast majority of all HIV infections occur through sexual intercourse. HIV can also be transmitted by infected blood or blood products, by the sharing of contaminated needles, and from an infected woman to her baby before birth, during delivery, or through breast-feeding. It is *not* spread through ordinary social contact.
3. **A worldwide problem.** More than 16 million adults and 1 million children had been infected with HIV by mid-1994 since the start of the pandemic, according to estimates by the World Health Organization. Around 4 million adults and children had developed AIDS. Although Africa has borne the brunt, no continent has been spared. HIV is now spreading fast in Asia and Latin America.
4. **Sexual transmission can be prevented.** Sexual transmission of HIV can be prevented by abstinence, fidelity between uninfected partners, and safer sex, which includes nonpenetrative sex and sex with condoms. Children need education about HIV prevention *before* they become sexually active. Everyone needs easy access to condoms in case of need.
5. **The family.** The concept of family need not be limited to ties of blood, marriage, sexual partnership, or adoption. Any group whose bonds are based on trust, mutual support, and a common destiny may be regarded as a family. So religious congregations, workers' associations, support groups of people with HIV/AIDS, gangs of street children, circles of drug injectors, collectives of sex workers, and networks of governmental, nongovernmental, and intergovernmental organizations may all be seen as families within the overarching family of humankind.
6. **The ripple effect on families.** Every day, around 6,000 people are newly infected with HIV. But several times this number will be newly *affected* by HIV every day through the impact on each infected individual's family and community.
7. **An extra threat in the 1990s.** Many families in the 1990s are disrupted by political upheaval, civil unrest, migration, and other factors. For millions of

them, HIV is an extra threat. If a breadwinner falls ill with AIDS, families face losses of income and sometimes food supply.

8. ***An additional burden for women.*** Nearly half of all newly infected adults are women. But, as women are the traditional care-givers, even *uninfected* women are affected by HIV when it enters a family. Women widowed by AIDS are often rejected and stripped of their belongings.

9. ***Children pay a growing price.*** Increasingly, children are paying the price of AIDS—either by being infected themselves or through the effect of AIDS on other family members. They may lose their parents and have to live on the streets if other relatives cannot or do not step in with support.

10. ***Families take care.*** All families, traditional and nontraditional, can help stop AIDS spreading by making sure that their members understand—and act on— the facts about HIV and safer behavior. And if one of their members *does* fall ill with AIDS, families are often the best source of compassionate care and support.

Source: World Health Organization. *World AIDS Day Newsletter,* 1994, No. 2.

COSTS OF COMMONLY USED AGENTS
AND LABORATORY TESTS FOR
AN HIV-INFECTED ADULT*

AGENT OR TEST	TRADE NAME	COMMON REGIMEN	COST/YR ($)
Antiretroviral treatment			
Zidovudine (AZT)	Retrovir	200 mg orally 3 times/day	3,490
Didanosine (ddI)	Videx	200 mg orally 2 times/day	2,160
Zalcitabine (ddC)	Hivid	0.75 mg orally 3 times/day	2,520
Stavudine (d4T)	Zerit	20 mg orally 2 times/day	2,730
Lamivudine (3TC)	Epivir	150 mg orally 2 times/day	2,800
Saquinavir	Invirase	600 mg orally 3 times/day	7,080
Ritonavir	Norvir	600 mg orally 2 times/day	8,120
Indinavir	Crixivan	800 mg orally 3 times/day	6,020
Nevirapine	Viramune	200 mg orally 2 times/day	3,020
Prophylaxis against			
Pneumocystis carinii			
pneumonia or			
toxoplasmosis			
TMP-SMX	Bactrim DS	160 mg of TMP and 800 mg	
		of SMX orally 3 times/wk	30
Dapsone		100 mg orally daily	70
Atovaquone	Mepron	750 mg orally 2 times/day	8,190
Prophylaxis against			
Mycobacterium			
avium complex			
Rifabutin	Mycobutin	300 mg orally daily	1,390
Clarithromycin	Biaxin	500 mg orally 2 times/day	2,380
Azithromycin	Zithromax	1200 mg orally every week	1,510
Antifungal prophylaxis			
or treatment			
Clotrimazole troche	Mycelex	10 mg orally 5 times/day	1,360
Nystatin suspension	Mycostatin	5 ml orally 4 times/day	2,660
Fluconazole	Diflucan	100 mg orally daily	2,510
Ketoconazole	Nizoral	200 mg orally daily	1,070
Itraconazole	Sporanox	200 mg orally daily	4,260

AGENT OR TEST	TRADE NAME	COMMON REGIMEN	COST/YR ($)
Prophylaxis against cytomegalovirus			
Ganciclovir, oral	Cytovene	1 g orally 3 times/day	17,080
Herpes simplex treatment			
Acyclovir	Zovirax	400 mg orally 2 times/day	1,610
Nutritional maintenance			
Food supplement	Ensure	1 8-oz can orally 3 times/day	1,610
	Ensure Plus	1 8-oz can orally 3 times/day	1,970
Dronabinol	Marinol	2.5 mg orally 2 times/day	1,820
Megestrol acetate	Megace	800 mg orally daily	3,290
Laboratory monitoring			
Complete blood count and chemistry profile[†]		4 times/yr	200
CD^4 cell count		4 times/yr	400
HIV viral-load measurement		4 times/yr	800

*Costs are based on average wholesale prices, rounded to the nearest $10, with prices for generic products used when available. TMP-SMX denotes trimethoprim-sulfamethoxazole.
†The chemistry profile includes measurements of electrolytes and renal and liver-function tests.
Table by Henry Feder Jr. M.D. University of Connecticut Health Center, Farmington, Ct. and Laura Milch, Pharm.D. Hartford Hospital, Hartford, Conn. as shown in *The New England Journal of Medicine* March 27, 1997:960.

LYMPHOCYTE FORMATION AND DESTINATIONS - NORMAL AND IN HIV INFECTIONS

During normal lymphocyte formation in polarized germinal centers (**PGC**), *as depicted on the right*, mesenchymal reticular cells (**MRC**) differentiate into large germinal center lymphocytes (**LGCL**), large germinal center macrophages (**LGCM**), endothelial cells (not shown) and follicular dendritic cells (**FDC**) which form a fine reticular stroma supporting all the former. Using influent substrates processed by the **LGCM** and **FDC**, the **LGCL** grow rapidly and divide by mitosis to generate a variety of soluble globulins, including nutritive globulins, antibodies, and lymphokines; along with many lymphocytes of smaller size. Large, medium-sized lymphocytes (**MSL**), and small cytoplasm-depleted lymphocytes (**SCDL**) are generated from the larger cells by means of mitosis, followed by progressive nuclear chromatin condensation and progresive loss of cytoplasm through shedding in the form of plasmalemma-encased globules which de-polymerize and disintegrate to produce sols rich in dissolved globulins. These soluble globulins flow into lymphatics or into blood circulation; whereas the **MSL** and **SCDL** accumulate around afferent arterioles and, then, actively migrate into circulation. From circulation the **SCDL** continue *emperipoletic* migration throughout the body; into most organs, tissues and many dividing tissue cells; and into most body secretions to help regulate cell growth and sustain nutritive, as well as immunologic *homeostasis.*

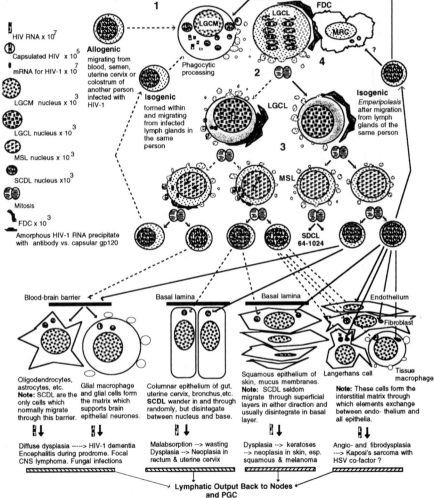

Fig. 1.5 **LYMPHOCYTE & HIV-1 KINETICS IN HIV-1 SICKNESS** (from *Lymphology* 1997;30:137-154)
Dashed arrows indicate pathways whereby HIV-1 RNA is phagocyte processed by **LGCM** and moved by the smaller cytoplasm-depleted progeny of **LGCL** to cause HIV-1 sickness. The differing positions of HIV-1 RNA 🔲 in nuclei are meant to indicate random and variable integration into lymphocyte proviral DNA. Presumably, the stability of HIV-1 capsular gp120 derived from the plasmalemma of **LGCL** is what prevents capsulated HIV-1, retroviral nucleoids and core antigens from circulating some 10-14 days after the **LGCL** in each **PGC** start producing appropriate antibodies ------ at least until the functional genes in **LGCL** are cumulatively altered by random HIV-1 RNA insertion. However, the random production of mutant proviral DNA coding for production of mutant HIV-1 RNA and mutant reverse transcriptase not expressed on cell surfaces is probably what keeps the disease process going almost *ad infinitum* in persons whose genes are receptive. Ultimately, a sprectrum of HIV-associated *homeostatic* failures may result in the lymph glands, brain, gut, uterus, skin and connective tissues, each failure partly being owing to the progressive random insertion of variably integrated HIV-1-RNA into critical genes, coupled with an insufficient supply of healthy *emperipoletic* SCDL.

A TECHNICAL DESCRIPTION OF HIV-1
SICKNESS AND TRANSMISSION

1   HIV-1 enters a receptive lymphocyte in a lymph gland germinal center. HIV
    is extremely small. If HIV retrovirus were the size of a tennis ball, the recep-
    tive germinal center lymphocyte would be the size of an aircraft carrier.

1–2 HIV-1 releases its RNA inside the lymphocyte. Single strand RNA randomly
    reverse transcribes into the lymphocyte DNA under the influence of virus-
    associated *reverse transcriptase*. This makes a template in double-stranded lym-
    phocyte DNA that can reproduce HIV-1 RNA and code for the production of
    HIV-1 from the cell when genes containing integrated HIV-1 RNA (called
    provirus) are appropriately stimulated.

2   The provirus infected cell containing integrated HIV-1 proviral DNA may
    shed myriad HIV-1 particles, along with other cell products, as shown in fig-
    ure 1.4 (see page 10).

2–3 The provirus infected lymphocyte divides into two daughter cells. Cell (left)
    may continue stimulated to produce virus particles. Daughter cell (right) may
    cease production, but continue to divide by mitosis 6–10 times over a period
    of 10–14 days to produce 64–1024 progeny containing randomly integrated
    provirus capable of HIV-1 production under the influence of co-factors and
    appropriate stimulation.

3   The most common end results during the course of HIV-1 sickness (HIVS)
    and progression into AIDS are as depicted below **3** in humans.

4   Pursuant to the signal studies of Barré-Sinoussi, Pantaleo, Embretson, Ho, Wei
    and many others during the last five years (Cf. JW Shields in *Lymphology* 1997;
    30:137–154), it has been established that the germinal centers of all lymph
    glands, including the lymph nodes in each region of the body, spleen, tonsils,
    adenoids, appendix, and small intestine become infected within 3 weeks after
    first exposure to HIV-1 provirus infected lymphocytes migrating from another
    infected person and within the body, as depicted at **4,** described in Chapter 3.

   Cogent here is that in order to sustain a state of constant nutritive and
    immunologic balance called *homeostasis* throughout the body, the germinal cen-
    ters normally give rise daily to untold billions of small lymphocytes which, in
    turn, migrate by unique motion called *emperipolesis* (Gr. EM-in; PERI-around;
    POLESIS-wandering) during which time they wander into and through other
    tissues, other cells, and other people. Many *emperipoletic* small lymphocytes nor-
    mally wander back to germinal centers to help regulate the system by

"biofeed-back" of DNA, which small lymphocytes contain in concentrations 5–10 times that in remaining body cells.

The crucial agent causing HIV Sickness/AIDS is a reverse-acting poison called *reverse transcriptase* (RT). RT is an enzyme that transcribes HIV-1 RNA at random into the DNA of dividing large germinal center lymphocytes (LGCL) to produce integrated provirus. HIV-1 RT acts during mitosis, when the LGCL genes double and the chromosomes split and then reassemble to produce two daughter cells.

Within the germinal centers of intestinal, splenic, lymph nodal, adenoidal and tonsilar lymphoid tissue, LGCL normally divide by mitosis 6–10 times over a period of 14 days to generate 64–1,024 progeny, each of which diminishes in size by nuclear condensation and shedding cytoplasm. Prior to mitosis, during the S-Phase of DNA synthesis, the LGCL reutilize significant quantities of DNA from migratory small cytoplasm-depleted lymphocytes (SCDL) recognized as being isologous and genetically compatible, along with DNA from allogenic SCDL whose DNA and cell products may not be recognized as foreign until they are serially processed by reticulo-endothelium, macrophages, and plasmacytes in lymph glands for 6–14 days. The combined results are that many HIV-1-infected allogenic SCDL from another genetically different person gaining access to a given person via blood, semen, uterine secretions, or milk are not recognized as foreign until it's too late, especially if many random insertions of retroviral RNA into isogenic lymphocyte DNA occur during the course of 6–10 LGCL and progenic mitoses. The side-effects may be:

- Random mutation in lymphocyte genes at the rate of 100–1,000 every 14 days with corresponding loss or alteration of critical genes normally operative.
- RT-driven random mutation in the provirus of HIV-1 produced by large lymphocytes with production of sufficient mutant strains every two weeks to offset effects of drugs, such as AZT and DDI, used to inhibit RT production (Cf. *Nature* 1997; 387:960–964)
- Emergence of sufficient mutant HIV-1 strains and mutant RT produced by such strains to negate the long-term beneficial effects of added protease inhibitors that have not yet been shown to eliminate HIV-1 provirus from circulating or tonsilar lymphocytes.
- Emergence of sufficient mutant HIV-1 strains to preclude development of good vaccines.
- Random cumulative mutations in lymphocyte genes over an average period of 10–12 years, ostensibly reflected progressively during prodrome, latency, PGL, ARC, ARD and in poorly defined, yet related "autoimmune" disorders—each and all depending on which and how many of ±100,000 potentially operative genes were singularly or cumulatively impaired by insertion of spurious retroviral nucleotide sequences into the precursors of 10–25 trillion lymphocytes normally migrating in the body to perpetually perform a variety of vital functions. As a result, the remaining 75 trillion cells in the body progressively suffer from malnutrition, disorderly growth, malfunction, tendencies to become malignant, and lack of resistance to infection.

REFERENCES

The material for this booklet has been drawn from many sources, the most important of which are the following:

All of the Conferences on AIDS.
HIV/AIDS Surveillance published by the CDC.
Morbidity and Mortality Weekly Report published by the CDC, 1981 to present.
Science, Nature, JAMA, Lancet, Lymphology, Blood, Journal of Infectious Diseases, and so forth.

For other information, see the following references:

Department of Health and Human Services. *Condoms and Sexually Transmitted Disease . . . Especially AIDS,* 1991 Publication FDA 90-4239.
Hatcher, Robert A., et al. *Contraceptive Technology, 1980–1995.* 15th rev. ed. New York: Irvington Publishers, 1996.
For an in-depth study of the first five years of the AIDS epidemic and its relations to the gay community, the government, politicalization, and the search for the virus read: *And the Band Played On* by Randy Shilts. New York: St. Martin Press, 1987.

For further information, look up the AIDS telephone hot line in your community. If you cannot find it, call the CDC National AIDS hotline: 1–800–342–2437 or 1–800–344–7432 (Spanish access) or 1–800–243–7889 (TTY, deaf access) to get a local number. Many cities and counties now have AIDS task forces. Some are run by medical societies, others by interested lay persons or by government officials.

GLOSSARY

ABSTINENCE   avoiding sexual relations.

ACQUIRED   a condition that is not inherited but is received from the environment.

ACQUIRED IMMUNITY   a continual state of immunity from a disease because the body has built up specific antibodies that prevent the disease from recurring.

AIDS (ACQUIRED IMMUNE DEFICIENCY SYNDROME). A currently incurable disease resulting in death.

AIDS-RELATED COMPLEX (ARC)   diseases of lymph glands and lymphocytes that may cause kidney and blood diseases or may, over time, transform into AIDS.

ANTIBODIES   complex proteins produced by the white blood cells that fight infections.

BISEXUAL   sexual desire for persons of both sexes.

CELL   a small form of life that forms the essential building blocks of your body.

CENTERS FOR DISEASE CONTROL (CDC)   the official U.S. government public health bureau responsible for tracking and preventing diseases in America.

CERVIX   the mouth or opening to the uterus that extends into the upper part of the vagina.

CONDOM (RUBBER)   a sheath of very fine latex or animal gut that fits over the penis and stops semen from entering the vagina during ejaculation. Animal gut condoms are not advised, as lymphocytes can migrate through the gut tissue.

DEFICIENCY   lacking or damaged in some way.

DIAPHRAGM   a dome-shaped cup of thin latex that, when properly placed in the vagina, blocks the entrance to the uterus.

ELISA   enzyme-linked immunoabsorbent assay—a test used to determine if a person's blood contains antibodies against the AIDS virus.

GERMINAL CENTERS   these are found in the lymph glands of mammals and coordinate the way food and oxygen are consumed to sustain life.

GLAND   an organized collection of cells having one or more specific functions in relation to the welfare of the body as a whole. For example, sweat glands produce sweat to help cool the body; reproductive glands (testes and ovaries) produce sperm and eggs, allowing reproduction.

HETEROSEXUAL   sexual desire for those of the opposite sex.

HIV-1   human immunodeficiency virus.

HOMOSEXUAL   sexual desire for those of the same sex.

HOST CELL   a healthy cell, the materials of which are used by a virus for nutrients and reproduction.

HYPERTENSION   high blood pressure.

IMMUNE   resistant to something.

IMMUNE SYSTEM a biochemical complex of cells in lymph glands and in blood that protect the body against infection, cancer, and other diseases.

INTRAVENOUS within a vein, as in injection of a substance directly into a person's bloodstream by using a needle placed into a vein.

KAPOSI'S SARCOMA (KS) an unusual cancer of blood and lymph vessels serving the skin, mucous membranes, and other glands in the body.

LESION a visible wound, sore, or rash.

LYMPH GLANDS organized collections of cells that produce fluids and products rich in a variety of globulens and lymphocytes essential to growth, regulation of bodily functions, and protection against disease.

LYMPHOCYTE a type of white blood cell formed in lymph glands and in bone marrow that produces antibodies and migrates throughout the body to nourish other cells, control growth, and fight infections.

LYMPHOMA cancers of the lymph glands.

MALIGNANT cancerous.

MONOGAMY the state of being married to or having sex with only one person.

MORTALITY RATE death rate.

NEUROLOGICAL having to do with the nervous system.

NUTRIENT food that the body needs to maintain health and normal growth.

ORAL THRUSH persistent white coating or spots inside the mouth or throat that may be accompanied by soreness and difficulty in swallowing.

PERSONS WITH AIDS (PWA) the preferred term for people having symptoms and signs of AIDS and positive ELISA tests.

PLACENTA the internal organ that develops in the uterus with pregnancy and through which the unborn child absorbs oxygen and nutrients and excretes wastes.

PRODROME premonitory symptom of disease.

PROMISCUOUS characterized by irresponsibility and lack of discrimination. Specifically, engaging in sexual relations or sharing needles with many different partners.

REMISSION partial or complete disappearance of symptoms, often only temporary.

SEMEN the secretion of the male reproductive organs that is ejaculated from the penis during orgasm and contains sperm cells and migrant lymphocytes.

SEXUALLY TRANSMITTED DISEASE (STD) any disease that is transmitted through sexual relations.

SHOOTING UP a slang term used to describe the injection of drugs into the veins.

SLIM DISEASE (WASTING DISEASE) is a disease with severe weight loss, body wasting, and weakness; sometimes associated with chronic diarrhea and persistent coughing.

SYNDROME distinct signs and symptoms that occur together to characterize a disease.

VIRUS a noncellular disease-producing organism that depends on a living cell for survival.

WHITE BLOOD CELL one element of the circulating lymph and blood systems. There are several types, all of which are essential to immunity.